Human Activity Recognition and Prediction

Yun Fu

Editor

Human Activity Recognition and Prediction

 Springer

Editor
Yun Fu
Northeastern University
Boston, Massachusetts, USA

ISBN 978-3-319-27002-9   ISBN 978-3-319-27004-3   (eBook)
DOI 10.1007/978-3-319-27004-3

Library of Congress Control Number: 2015959271

Springer Cham Heidelberg New York Dordrecht London
© Springer International Publishing Switzerland 2016

Printed on acid-free paper

Springer International Publishing AG Switzerland is part of Springer Science+Business Media (www.springer.com)

Preface

Automatic human activity sensing has drawn much attention in the field of video analysis technology due to the growing demands from many applications, such as surveillance environments, entertainments, and healthcare systems. Human activity recognition and prediction is closely related to other computer vision tasks such as human gesture analysis, gait recognition, and event recognition. Very recently, the US government funded many major research projects on this topic; in industry, commercial products such as the Microsoft's Kinect are good examples that make use of human action recognition techniques. Many commercialized surveillance systems seek to develop and exploit video-based detection, tracking and activity recognition of persons, and vehicles in order to infer their threat potential and provide automated alerts.

This book focuses on the recognition, prediction of individual activities and interactions from videos that usually involves several people. This provides a unique view of: human activity recognition, especially fine-grained human activity structure learning, human interaction recognition, RGB-D data-based recognition temporal decomposition, and casually learning in unconstrained human activity videos. These techniques will significantly advance existing methodologies of video content understanding by taking advantage of activity recognition. As a professional reference and research monograph, this book includes several key chapters covering multiple emerging topics in this new field. It links multiple popular research fields in computer vision, machine learning, human-centered computing, human-computer interaction, image classification, and pattern recognition. Contributed by top experts and practitioners of the Synergetic Media Learning (SMILE) Lab at Northeastern University, these chapters complement each other from different angles and compose a solid overview of the human activity recognition and prediction techniques. Well-balanced contents and cross-domain knowledge for both methodology and real-world applications will benefit readers from different level of expertise in broad fields of science and engineering.

There are in total eight chapters included in this book. The book will first give an overview of recent studies for activity recognition and prediction in the "Introduction" chapter, such as objectives, challenges, representations, classifiers, and datasets, and will then discuss several interesting topics of these fields in details. Chapter 2 addresses the challenges of action recognition through interactions and proposes semantic descriptions, namely, "interactive phrases," to better describe complex interactions, and discriminative spatiotemporal patches to provide cleaner features for interaction recognition. Chapter 3 presents a sparse tensor subspace learning method to select variables for high-dimensional visual representation of action videos, which not only keeps the original structure of action but also avoids the "curse of dimensionality." Chapter 4 introduces a multiple mode-driven discriminant analysis (MDA) in tensor subspace action recognition, which preserves both discrete and continuous distribution information of action videos in lower dimensional spaces to boost discriminant power. Chapter 5 presents a transfer learning method that can transfer the knowledge from depth information of the RGB-D data to the RGB data and use the additional source information to recognize human actions in RGB videos. In particular, a novel cross-modality regularizer is introduced that plays an important role in finding the correlation between RGB and depth modalities, allowing more depth information from the source database to be transferred to that of the target. Chapter 6 introduces a multiple temporal scale support vector machine for early classification of unfinished actions; and a convex learning formulation is proposed to consider the essence of the progressively arriving action data. Chapter 7 discusses an approach for predicting long duration complex activity by discovering the causal relationships between constituent actions and predictable characteristics of the activities. Chapter 8 introduces an approach to early classify human activities represented by multivariate time series data, where the spatial structure of activities is encoded by the dimensions of predefined human body model and the temporal structure of activities are modeled by temporal dynamics and sequential cue.

This book can be used by broad groups of readers, such as professional researchers, graduate students, and university faculties, especially those in the background of computer science and computer engineering. I would like to sincerely thank all the SMILE Lab contributors of this book for presenting their most recent research advances in an easily accessible manner. I would also sincerely thank editors Mary E. James and Rebecca R. Hytowitz from Springer for strong support to this book project.

Boston, MA, USA   Yun Fu

Contents

Chapter 1
Introduction

Yu Kong and Yun Fu

The goal of human action recognition is to predict the label of the action of an individual or a group of people from a video observation. This interesting topic is inspired by a number of useful real-world applications, such as visual surveillance, video understanding, etc. Considering that in a large square, an online visual surveillance for understanding a group of people's action will be of great importance for public security; an automatic video understanding system will be very effective to label millions of online videos.

However, in many real-world scenarios (e.g., vehicle accident and criminal activity), intelligent systems do not have the luxury of waiting for the entire video before having to react to the action contained in it. For example, being able to predict a dangerous driving situation before it occurs; opposed to recognizing it thereafter. This task is referred to as action prediction where approaches that can recognize progressively observed video segments, different to action recognition approaches that expect to see the entire set of action dynamics extracted from a full video.

Although conventional color videos contain rich motion and appearance information, they do not provide structural information of the entire scene. In other words, machines cannot tell which object in a video is more closer to the camera and which is more far away. Due to the recent advent of the cost-effective Kinect sensors, action recognition from RGB-D cameras is receiving increasing interests in computer vision community. Compared with conventional RGB cameras, Kinect

Y. Kong (✉)
Department of Electrical and Computer Engineering, Northeastern University,
360 Huntington Avenue, Boston, MA 02115, USA
e-mail: yukong@ece.neu.edu

Y. Fu
Department of Electrical and Computer Engineering and College of Computer and Information
Science (Affiliated), Northeastern University, 360 Huntington Avenue, Boston, MA 02115, USA
e-mail: yunfu@ece.neu.edu

© Springer International Publishing Switzerland 2016
Y. Fu (ed.), *Human Activity Recognition and Prediction*,
DOI 10.1007/978-3-319-27004-3_1

sensors provide depth information, which captures 3D structural information of the entire scene. The 3D structural information can be used to facilitate the recognition task by simplifying intra-class motion variation and removing cluttered background noise.

In this chapter, we will first review recent studies in action recognition and prediction that consist of action representations, action classifiers, and action predictors. Approaches for recognizing RGB-D videos will also be discussed. We then will describe several popular action recognition datasets, including ones with individual actions, group actions, unconstrained datasets, and RGB-D action datasets. Some of existing studies [106, 107] aim at learning actions from static images, which is not the focus of this book.

1 Challenges of Human Activity Recognition and Prediction

Despite significant progress has been made in human activity recognition and prediction, the most advanced algorithms still misclassify action videos due to several major challenges in this task.

1.1 Intra- and Inter-Class Variations

As we all know, people behave differently for the same actions. For a given semantic meaningful activity, for example, "running," a person can run fast, slow, or even jump and run. That is to say, one activity category may contain multiple different styles of human motion. In addition, videos in the same action can be captured from various viewpoints. They can be taken in front of the human subject, on the side of the subject, or even on top of the subject. Furthermore, different people may show different poses in executing the same activity. All these factors will result in large intra-class appearance and pose variations, which confuse a lot of existing activity recognition algorithms. These variations will be even larger on real-world activity datasets. This triggers the investigation of more advanced activity recognition algorithms that can be deployed in real-world.

Furthermore, similarities exist in different activity categories. For instance, "running" and "walking" involve similar human motion patterns. These similarities would also be challenging to differentiate for intelligent machines, and consequently contribute to misclassifications.

In order to minimize intra-class variations and maximize inter-class variations, lots of effort have been made to design discriminative activity features. In some recent studies, researchers attempt to learn discriminative features using deep learning techniques, in order to better describe complex human activities.

1.2 Cluttered Background and Camera Motion

It is interesting to see that a number of human activity recognition algorithms work very well in indoor environment but fail in outdoor environment. This is mainly due to the background noise. In real-world, existing activity features will also encode background noise and thus degrade the recognition performance.

Camera motion is another factor that should be considered in real-world applications. Due to significant camera motion, activity features cannot be accurately extracted. In order to better extract activity features, camera motion should be modeled and compensated.

Other environment-related issues such as illumination conditions, viewpoint changes, dynamic background will also be the challenges that prohibit activity recognition algorithms from being used in the real-world.

1.3 Insufficient Data

A robust and effective intelligent system normally requires large amount of training data. Even though existing activity recognition systems [31, 55, 60] have shown impressive performance on small-scale datasets in laboratory settings, it is really challenging to generalize them to real-world applications due to the unavailability of large action datasets. Most of existing activity datasets only contain about thousands of videos (e.g., UCF Sports dataset [70]), which would not be enough to train a sophisticated activity recognition algorithm.

Furthermore, it is labor intensive to annotate a large dataset. Some sophisticated algorithms [10, 38, 67] require the bounding boxes of the person of interest to be annotated. This is infeasible on large dataset. However, it is possible that some of the data annotations are available, which would result in a training setting with a mixture of labeled data and unlabeled data. Therefore, it is imperative to design an activity recognition algorithm that can learn activities from both labeled data and unlabeled data.

2 Video Representations

2.1 Global Features

Global features capture holistic information of the entire scene, including the appearance feature, geometric structure, motion, etc. However, global features are usually sensitive to noise and clutter background since they also bring in some information that is irrelevant to activity recognition. Global features such as histograms of oriented histograms (HOG) [11] and optical flow [15, 56] have been successfully applied in activity recognition.

HOG captures object appearance and shape information and has shown to be invariant to illumination, shadowing, etc. It computes gradient in local image cells and then aggregates the gradients by weighted voting into spatial and orientation cells. Though popularly used in object detection, recent work uses a track of HOG features for activity recognition [7, 27, 34], or adapts HOG to capture appearance information within local spatiotemporal regions [31, 46, 90].

Optical flow [56] captures motion fields on successive frames. Under the assumption that illumination conditions do not change on the frames, optical flow computes the motion in horizontal and vertical axis. To better represent optical flow fields, Efros et al. [15] further split the horizontal and vertical fields into positive and negative channels, and concatenate the four channels to characterize human motion in a distance. Optical flow feature is further aggregated in histograms, called histograms of optical flow (HOF), and combined with HOG features to represent complex human activities [31, 46, 90]. Researchers also compute gradients over optical flow fields and build so-called the motion boundary histograms (MBH) for describing trajectories [90].

2.2 Local Features

Local features such as spatial-temporal interest points by [5, 14, 31, 45] are very popularly used in recent studies in action recognition due to their robustness to translation, appearance variation, etc. Different from global features, local features describe local motion of a person in space-time regions. These regions are detected since the motion information within the regions are more informative and salient than surrounding areas. After detection, the regions are described by extracting features in the regions.

Laptev [44] extended the 2D Harris corner detector to space-time domain, and validated its effectiveness in outdoor sequences. Dollar et al. [14] and Bregonzio et al. [5] detected spatial-temporal interest points using Gabor filters. Spatiotemporal interest points can also be detected by using the spatiotemporal Hessian matrix [101]. Other detection algorithms detect spatiotemporal interest points by extending their counterparts of 2D detectors to spatiotemporal domains, such as 3D SIFT [76], HOG3D [31], local trinary patterns [108], etc. Several descriptors have been proposed to describe the motion and appearance information within the small region of the detected interest points. For example, gradient, histogram of oriented gradient, histogram of optical flow, and motion boundary of histogram.

However, spatiotemporal interest points only capture information within a short temporal duration and cannot capture long-term duration information. It would be better to track these interest points and describe their changes of motion properties. Feature trajectory is a straightforward way of capturing such long-duration information [68, 91, 92]. To obtain features for trajectories, interest points are first detected and tracked using KLT tracker [56] or pairwise SIFT matching [82].

The point-level context is captured in [82] by averaging all trajectory features. Trajectories are described by a concatenation of HOG, HOF, and MBH features [24, 92, 96], intra- and inter-trajectory descriptors [82], or HOG/HOF and averaged descriptors [68]. In order to reduce the side effect of camera motion, [89, 98] find correspondences between two frames first and then use RANSAC to estimate the homography.

All the above methods rely on hand-crafted features, which require expensive human labor to fine-tune parameters. Recent work also shows that action features can also be learned using deep learning techniques [26].

2.3   Deeply Learned Features

Although great success has been achieved by global and local features, these hand-crafted features require heavy human labor and require domain expert knowledge to design the feature extraction framework. In addition, they normally do not generalize very well on large datasets.

In recent years, feature learning using deep learning techniques has been receiving increasing attention due to their ability of designing power features that can be generalized very well. In addition, these learning techniques are able to perform semi-supervised feature abstraction, allowing us to use side information, and hierarchical feature abstraction. Deep learning is a branch of machine learning, which aims at modeling high-level abstractions in data by complex structured model architectures composed of multiple non-linear transformations.

Action features learned using deep learning techniques have been popularly investigated in recent years [2, 3, 21, 25, 26, 29, 47, 66, 79, 83, 86, 97, 99, 104]. Specifically, a 3D convolution operation in the convolutional neural network (CNN) was proposed in [25, 26] to learn discriminative features from videos. Simonyan and Zisserman [79] proposed a convolutional network that filters videos in two separate streams, i.e. spatial and temporal, and then combines using a late fusion. Wang et al. [97] treat an action video as a set of cubic-like temporal segments, and discover temporal structures using CNN. Taylor et al. [86] generalized the gated restricted Boltzmann machine to extracting features from sequences. Karpathy et al. [29] complied a very large Sports-1M dataset for video classification, and conducted extensive experiments of CNNs. An extension of independent component analysis was proposed in [47] that employs the idea of convolution and stacking in order to scale up to big datasets. The sparse Autoencoder is employed to incrementally learning human activities in streaming videos [21]. An architecture of hierarchical depth motion maps (HDMM) + 3 Channel Convolutional Neural Network (3ConvNets) is proposed in [99] for recognizing human actions from depth sequences.

3 Action Classifiers

3.1 Individual Action Classifiers

A considerable amount of previous work in human action recognition focuses on recognizing the actions of a single person in videos [1, 18, 40, 60, 77].

3.1.1 Space-Time Approaches

Conventional space-time action recognition methods mainly focus on learning motion patterns from holistic or local space-time motion patterns. A global space-time representation learning algorithm was proposed in [4, 18]. The proposed method utilizes the properties of the solution to the Poisson equation to extract space-time features such as local space-time saliency, action dynamics, shape structure, and orientation. A holistic body shape feature was also used in [53] to characterize the shape of human body during the execution of actions. Recent work in [67] showed that representative key poses can be learned to better represent human actions. This method discards a number of non-informative poses in a temporal sequence, and builds a more compact pose sequence for classification. Nevertheless, these global feature-based approaches are sensitive to background noise and generally do not perform well on challenging datasets.

Local space-time approaches are receiving more attention recently. Spatiotemporal interest points-based approaches [14, 31, 44, 75] were proposed to detect local salient regions first, and then build a so-called dictionary to quantize the features extracted from the local regions. An action video can be finally represented by a histogram of these quantized features. Although these local features only capture motion of body parts and fail to describe the holistic motion patterns, they have been shown to be insensitive to appearance and pose variations, and achieved promising results in activity recognition. A potential drawback of the above action representation is that it does not consider structural information of the interest points. To overcome this problem, [73] measured the spatiotemporal relationships of interest points, and builds a new descriptor that embeds structural information.

3.1.2 Sequential Approaches

Another line of work captures temporal evolutions of appearance or pose using sequential state models [60, 78, 85, 95]. These approaches treat a video as a composition of temporal segments. However, they do not model temporal action evolution with respect to observation ratios. Therefore, they cannot characterize partially observed actions and are unsuitable for prediction. In addition, recognizing human actions from a set of key frames [67, 87] has also been investigated in previous studies. These approaches learn representative key frames to encode long duration action sequences, and thus reduce redundant features in classification. In some cases, a long temporal sequence may contain a series of inhomogeneous action units. Previous work [23, 51, 78] studied the problem of segmenting such long sequences into multiple semantic coherent units by considering motion relationships between successive frames.

3.1.3 Hierarchical Approaches

Bag-of-words models have shown to be robust to background noise but may not be expressive enough to describe actions in the presence of large appearance and pose variations. In addition, they may not well represent actions due to the large semantic gap between low-level features and high-level actions. To address these two problems, hierarchical approaches [10, 38, 55, 88] were proposed to learn an additional layer of representations, and expect to better abstract the low-level features for classification.

Hierarchical approaches learn mid-level features from low-level features, which are then used in the recognition task. The learned mid-level features can be considered as knowledge discovered from the same database used for training or being specified by experts. Recently, semantic descriptions or attributes are popularly investigated in activity recognition. These semantics are defined and further introduced into the activity classifiers in order to characterize complex human actions [36, 38, 55]. These approaches have shown superior performance since they introduce human knowledge into classification models.

Other hierarchical approaches such as [67, 87] select key poses from observed frames, which also learn better activity representations during model learning.

3.2 Classifiers for Group Actions

3.2.1 Human–Human Interactions

The recognition of human–human interactions from videos has been extensively investigated in recent studies. In most of the previous studies [46, 54, 57, 73, 109], interactions are recognized in the same way as single-person action recognition. Specifically, interactions are represented as a motion descriptor including all the people in a video, and then an action classifier such as linear support vector machine is adopted to classify interactions. Although reasonable performance can be achieved, these approaches do not utilize the intrinsic properties of interactions, such as co-occurrence information between interacting people. In addition, they treat people as a single entity and do not extract the motion of each person from the group. Thus, their methods could not give the action label of each interacting person in the video simultaneously.

Exploiting rich contextual information in human interactions can help achieve more accurate and robust recognition results. In [62], human interactions were recognized using motion co-occurrence, which is achieved by coupling motion state of one person with the other interacting person. Spatiotemporal crowd context is captured in [10] to recognize human interactions. Human pose, velocity, and spatiotemporal distribution of individuals are applied to represent the crowd context information. They further developed a system that can simultaneously track multiple people and recognize their interactions [8]. They capture co-occurrence between interaction and atomic action, and interaction and collective action for interaction recognition. The method in [43] studied the collective activity recognition problem using crowd context. People in a collective activity have no close physical contact with each other and perform similar action, e.g., "crossing the road," "talking," or "waiting." Individuals in interactions can also be represented by a set of key poses, and their spatial and temporal relationships can be captured for recognition [87]. Odashima et al. [61] proposed the Contextual Spatial Pyramid to detect the action of multiple people. In [36, 38], a semantic description-based approach was proposed to better represent complex human interactions. A patch-aware model was proposed in [33] to learn discriminative patches for interaction recognition. In these studies, the contextual information is easy to capture since the human body size in their datasets is not very small and thus a visual tracker for human body is performed well. However, in [48, 58, 59], the size of human body can often be very small, making them difficult or even impossible to track using tracker. Consequently, the contextual information between interacting entities is difficult to extract.

Human interactions, e.g., "hug," "push," and "high-five," usually involve frequent close physical contact. Ryoo and Aggarwal [72] utilized body part tracker to extract each individual in videos and then applied context-free grammar to describe spatial and temporal relationships between people. Perez et al. [65] adopted a human detector to extract individual in videos. They investigated interaction recognition between two people in realistic scenarios. In their work, spatial relationships between individuals are captured using the structured learning technique. However, the ambiguities in feature-to-person assignments during close physical contact remain a problem.

3.2.2 Human–Object Interactions

Contextual information has also been effectively exploited in human–object interactions and object–object interactions. Human–object interactions in videos and images are performed in [19] by fusing context from object reaction, object class, and manipulation motion. They incorporated rich context derived from object class, object reaction, and manipulation motion into Bayesian models for recognizing human–object interaction from videos and static images.

Mutual context of objects and human poses was explored by Yao and Fei-Fei [105]. Their work on human–object interaction recognition showed that using mutual context, solving human pose estimation and object detection problems simultaneously can greatly benefit each other. In [17], temporal and casual relationships between object events are represented by the dynamically interconnected multiple hidden Markov models, and are used for recognizing group actions involving multiple objects. A contextual model is proposed to capture geometric configurations of objects and human pose in contextual models for recognizing human–object interactions (e.g., tennis-serve and tennis-forehand) [13]. The relationships between human and object are modeled by joint actor–object states [16]. A discriminative model is proposed to capture the spatial relationships (e.g., above, below) between objects for multi-object detection [12].

3.3   Classifiers for RGB-D Videos

Recently, due to the advent of the cost-effective Kinect sensor, researchers pay a lot of attentions to recognizing actions from RGB-D data [20, 52, 63, 93, 94, 102]. Compared with conventional RGB data, the additional depth information allows us to capture 3D structural information, which is very useful in removing background and simplifying intra-class motion variations. Compared with these methods, depth information is not available in the target training and testing database in this work. [27] used an additional RGB-D data as the source database, and learn the correlations between RGB data and depth data. The learned correlation knowledge is then transferred to the target database, in which the depth information does not exist. With the learned depth information, the performance on the target RGB data can be improved comparing with the performance when only the RGB data in the target data are used.

3.4   Action Predictors

Most of existing action recognition methods [67, 87, 106, 107] were designed for recognizing complete actions, assuming the action in each testing video has been fully executed. This makes these approaches unsuitable for predicting action labels in partial videos.

Action prediction approaches aim at recognizing unfinished action videos. Ryoo [71] proposed the integral bag-of-words (IBoW) and dynamic bag-of-words (DBoW) approaches for action prediction. The action model of each progress level is computed by averaging features of a particular progress level in the same category. However, the learned model may not be representative if the action videos of the same class have large appearance variations, and it is sensitive to outliers. To overcome these two problems, Cao et al. [6] built action models by learning feature bases using sparse coding and used the reconstruction error in the likelihood computation. Li et al. [50] explored long-duration action prediction problem. However, their work detects segments by motion velocity peaks, which may not be applicable on complex outdoor datasets. Compared with [6, 35, 37, 50, 71], we incorporate an important prior knowledge that informative action information is increasing when new observations are available. In addition, the methods in [35, 37] model label consistency of segments, which is not presented in their methods.

Additionally, an early event detector [22] was proposed to localize the starting and ending frames of an incomplete event. Activity forecasting, which aims at reasoning about the preferred path for people given a destination in a scene, has been investigated in [30].

4 Datasets

An overview of popularly used datasets for activity recognition and prediction is given in the following. These datasets differ in the number of human subjects, background noise, appearance and pose variations, camera motion, etc., and have been widely used for the comparison of various algorithms.

4.1 Individual Action Video Datasets

Weizmann Dataset The Weizmann human action dataset [4] is a popular video dataset for human action recognition. The dataset contains ten action classes such as "walking," "jogging," "waving" performed by nine different subjects, to provide a total of 90 video sequences. The videos are taken with a static camera under a simple background.

KTH Dataset The KTH dataset [75] is a challenging dataset for human action recognition which consists of six types of human actions (boxing, hand clapping, hand waving, jogging, running, and walking) performed several times by 25 different subjects in four scenarios (outdoors, outdoors with scale variation, outdoors with different clothes, and indoors). There are 599 action videos in total in the KTH dataset.

INRIA XMAS Multi-View Dataset This dataset is a multi-view dataset for view-invariant human action recognition [100]. It contains videos captured from five views including a top-view camera. The dataset consists of 13 actions, each of which is performed 3 times by 10 actors.

Multicamera Human Action Video Dataset The multicamera human action video dataset [80] is composed of 17 action categories performed by 14 actors. Videos in this dataset are captured by eight cameras around the human subject in a constrained environment. Example actions are "walk and turn back," "run and stop," "punch," and "draw graffiti," etc.

IXMAS Actions Dataset This dataset is also a multi-view action dataset [100]. It consists of 11 actions, providing a total of 1148 videos captured by five cameras. 395 out of 1148 videos do not have occlusions and 698 videos contain objects that are partially occluded the actors.

MSR Action Dataset I MSR action dataset I [111] was captured with clutter and moving background. It consists of 63 videos, performed by ten subjects, in both indoor and outdoor environments. Each sequence contains multiple types of actions.

MSR Action Dataset II It is an extended version of MSR action dataset I [112]. Fifty-four videos captured in a crowded environment are provided, where each video contains multiple actions. Three action categories, hand waving, hand clapping, and boxing, are included in this dataset.

Example images of these individual action datasets can be found in Fig. 1.1.

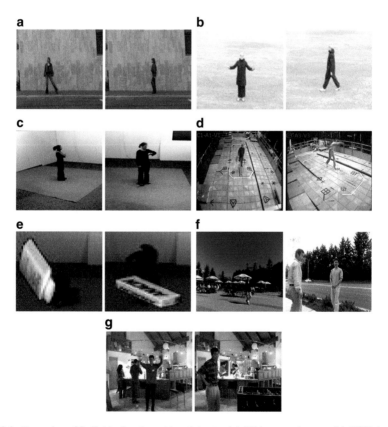

Fig. 1.1 Examples of individual action video datasets. (**a**) Weizmann dataset. (**b**) KTH dataset. (**c**) INRIA XMAS multi-view dataset. (**d**) Multicamera human action video dataset. (**e**) IXMAS actions dataset. (**f**) MSR action dataset I. (**g**) MSR action dataset II

4.2 Group Action Datasets

UT-Interaction Dataset UT-interaction dataset [74] is comprised of two sets with different environments. Each set consists of six types of human interactions: handshake, hug, kick, point, punch, and push. Each type of interactions contains 10 videos, to provide 60 videos in total. Videos are captured at different scales and illumination conditions. Moreover, some irrelevant pedestrians are present in the videos.

BIT-Interaction Dataset BIT-interaction dataset [36] consists of eight classes of human interactions (bow, boxing, handshake, high-five, hug, kick, pat, and push), with 50 videos per class. Videos are captured in realistic scenes with cluttered backgrounds, partial occluded body parts, moving objects, and variations in subject appearance, scale, illumination condition, and viewpoint.

TV-Interaction Dataset The TV-interaction dataset [64] contains 300 video clips with human interactions. These videos are categorized into four interaction categories: handshake, high five, hug, and kiss. The dataset is annotated with the upper body of people, discrete head orientation, and interaction label.

Collective Activity Dataset Collective activity dataset [9] was used for multi-person interaction recognition. This dataset consists of five group activities, crossing, talking, waiting, walking, and queueing. Forty-four short video clips of group action categories were recorded. Every tenth frame in every video is manually labeled with pose, activity, and bounding box.

LIRIS Human Activities Dataset The LIRIS human activities dataset was used in ICPR HARL 2012 contest. There are ten action classes in the dataset: discussion of two or several people, a person gives an item to a second person, handshaking of two people, a person types on a keyboard, a person talks on a telephone, etc.
Example images of these group action datasets are illustrated in Fig. 1.2.

4.3 Unconstrained Datasets

UCF Sports Action Dataset The UCF sports action dataset [70] contains a set of sports actions that are typically featured on television channels. It contains ten actions, providing a total of 150 videos. The actions are diving, golf swing, kicking, lifting, riding horse, running, skate boarding, swing-bench, swing-side, and walking.

Olympic Sports Dataset The Olympic sports dataset [60] contains sports videos of athletes in Olympic games. These videos were collected from YouTube and were annotated with the help of Amazon Mechanical Turk. The dataset has 16 sport actions, such as high-jump, triple-jump, shot-put, etc.

Fig. 1.2 Example images of group action datasets. (**a**) UT-interaction dataset. (**b**) BIT-interaction dataset. (**c**) TV-interaction dataset. (**d**) Collective activity dataset. (**e**) LIRIS human activities dataset

Hollywood Dataset The Hollywood dataset [46] contains video clips with human actions from 32 movies. There are eight action classes in the dataset, such as answer phone, sit up, and kiss.

Hollywood 2 Dataset The Hollywood 2 dataset [57] provides 12 classes of human actions (e.g., answer phone, eat, and drive a car) and ten classes of scenes (e.g., in car, in office, and in restaurant). There are in total of 3669 video clips from 69 movies. The total length of these videos is approximately 20.1 h.

UCF11 Dataset UCF11 action dataset [28] contains 11 action categories such as basketball shooting, biking, tennis swinging, etc. The videos are compiled from YouTube, and are very challenging due to large variations in camera motion, object appearance and scale, viewpoint, cluttered background, etc. For each category, the videos are grouped into 25 groups with at least four videos in it.

UCF50 Dataset The UCF50 dataset [69] contains 50 action types, such as biking, push-ups, mixing batter, etc. It is an extension of the UCF11 dataset. Videos in this dataset are realistic videos downloaded from YouTube. This dataset is very challenging due to large variations in camera motion, object appearance and pose, object scale, viewpoint, cluttered background, illumination conditions, etc.

UCF101 Dataset The UCF101 dataset [81] is an extended version of UCF50 dataset. It comprises of realistic videos collected from YouTube. It contains 101 action categories, with 13,320 videos in total. UCF101 gives the largest diversity in terms of actions and with the presence of large variations in camera motion, object appearance and pose, object scale, viewpoint, cluttered background, illumination conditions, etc.

HMDB51 Dataset The HMDB51 dataset [41] contains a total of about 7000 video clips distributed in a large set of 51 action categories. Each category contains a minimum of 101 video clips. In addition to the label of the action category, each clip is annotated with an action label as well as a meta-label describing the property of the clip, such as visible body parts, camera motion, camera viewpoint, number of people involved in the action, and video quality.

ASLAN Dataset The ASLAN dataset [32] is a full testing protocol for studying action similarity from videos. It contains 3697 action videos sampled from 1571 YouTube videos, distributed in 432 action categories. The average number of samples per class is 8.5. Three hundred and sixteen classes contain over one sample, meaning that the other classes only contain one sample, making the dataset more challenging.

Sports-1M Dataset The Sports-1M dataset [29] contains 1,133,158 video URLs, which have been annotated automatically with 487 labels. It is one of the largest video datasets. Very diverse sports videos are included in this dataset, such as tail, Shaolin kung fu, wing chun, etc. The dataset is extremely challenging due to very large appearance and pose variations, significant camera motion, noisy background motion, etc.

Figure 1.3 shows example figures of these unconstrained datasets.

4.4 RGB-D Action Video Datasets

MSR Gesture Dataset MSR-Gesture3D dataset [42] is a hand gesture dataset containing 336 depth sequences captured by a depth camera. There are 12 categories of hand gestures in the dataset, "bathroom," "blue," "finish," "green," "hungry," "milk," "past," "pig," "store," "where," "j," and "z." This is a challenging dataset due to the self-occlusion issue and visually similarity.

MSR Action3D Dataset MSR-Action3D dataset [49] consists of 20 classes of human actions: "bend," "draw circle," "draw tick," "draw x," "forward kick," "forward punch," "golf swing," "hammer," "hand catch," "hand clap," "high arm wave," "high throw," "horizontal arm wave," "jogging," "pick up & throw," "side boxing," "side kick," "tennis serve," "tennis swing," and "two hand wave." A total of 567 depth videos are contained in the dataset which are captured using a depth camera.

MSR Action Pairs Dataset In the MSR pair action database [63], there are six pairs of actions: pick up a box/put down a box; lift a box/place a box, push a chair/pull a chair, wear a hat/take off a hat, put on a backpack/takeoff a backpack, stick a poster/remove a poster. There are total 360 RGB samples and 360 depth

Fig. 1.3 Examples of unconstrained datasets. (**a**) UCF sports action dataset. (**b**) Olympic sports dataset. (**c**) Hollywood dataset. (**d**) Hollywood 2 dataset. (**e**) UCF11 dataset. (**f**) UCF50 dataset. (**g**) UCF101 dataset. (**h**) HMDB51 dataset. (**i**) ASLAN dataset. (**j**) Sports-1M dataset

action samples. Each action is performed three times by ten different subjects, where the actions of the first five subjects are used for testing and the rest for training.

MSR Daily Activity Dataset In the MSR daily action database [93], there are 16 categories of actions: drink, eat, read book, call cell phone, write on a paper, use laptop, use vacuum cleaner, cheer up, sit still, toss paper, play game, lie down on sofa, walk, play guitar, stand up, sit down. All these actions are performed by 10 subjects, each performs every action twice. There are 320 RGB samples and 320 depth samples available.

3D Online Action Dataset The 3D online action dataset [110] was designed for three evaluation tasks: same-environment action recognition, cross-environment action recognition, and continuous action recognition. The dataset contains human action or human–object interaction videos captured from RGB-D sensors. It contains seven action categories, such as drinking, eating, using laptop, and reading cell phone.

CAD-60 Dataset The CAD-60 dataset [84] comprises of RGB-D action videos that are captured using the Kinect sensor. There are 60 action videos in total, which are performed by four subjects in five different environments including office, kitchen, bedroom, bathroom, and living room. The dataset consists of 12 action types, such as rinsing mouth, talking on the phone, cooking, and writing on whiteboard. Tracked skeletons, RGB images, and depth images are provided in the dataset.

CAD-120 Dataset The CAD-120 dataset [39] comprises of 120 RGB-D action videos of long daily activities. It is also captured using the Kinect sensor. Action videos are performed by four subjects. The dataset consists of ten action types, such as rinsing mouth, talking on the phone, cooking, and writing on whiteboard. Tracked skeletons, RGB images, and depth images are provided in the dataset.

UTKinect-Action Dataset The UTKinect-action dataset [103] was captured by a Kinect device. There are ten high-level action categories contained in the dataset, such as making cereal, taking medicine, stacking objects, and unstacking objects. Each high-level action can be comprised of ten sub-activities such as reaching, moving, eating, and opening. Twelve object affordable labels are also annotated in the dataset, including pourable, drinkable, and openable.

Figure 1.4 displays example figures of these RGB-D action datasets.

4.5 Summary

Activity recognition is an important application in the computer vision community due to broad applications, e.g., visual surveillance. However, it is really challenging in real-world scenarios due to many factors, such as inter-class variations and cluttered background, which have been discussed in this chapter. Activities can be represented by different kinds of features, for example, trajectory features, spatiotemporal features, etc. These features are further classified by activity classifiers to obtain the activity labels. In this chapter, popularly used activity features are reviewed and a series of recent activity classifiers are also discussed. Various activity classifiers can be adopted for different classification goals, for instance, action recognition, human interaction recognition, action prediction, etc. We have also reviewed a group of activity datasets, including individual action video datasets, group action datasets, unconstrained datasets, and RGB-D action video datasets.

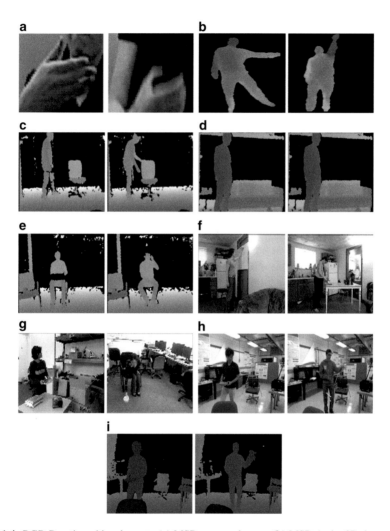

Fig. 1.4 RGB-D action video datasets. (**a**) MSR gesture dataset. (**b**) MSR Action3D dataset. (**c**) MSR action pairs dataset. (**d**) MSR daily activity dataset. (**e**) 3D online action dataset. (**f**) CAD-60 dataset. (**g**) CAD-120 dataset. (**h**) UTKinect-action dataset. (**i**) UTKinect-action dataset (depth images)

References

1. Aggarwal, J., Ryoo, M.S.: Human activity analysis: a review. ACM Comput. Surv. **43**(3), pp. 16:1–16:43 (2011)
2. Baccouche, M., Mamalet, F., Wolf, C., Garcia, C., Baskurt, A.: Sequential deep learning for human action recognition. In: Human Behavior Understanding, Springer Berlin Heidelberg, (2011)

3. Bengio, Y., Courville, A., Vincent, P.: Representation learning: a review and new perspectives. IEEE Trans. Pattern Anal. Mach. Intell. vol. 35, pp. 1798–1828 (2013)
4. Blank, M., Gorelick, L., Shechtman, E., Irani, M., Basri, R.: Actions as space-time shapes. In: International Conference on Computer Vision, vol. 2, pp. 1395–1402. IEEE, New York (2005)
5. Bregonzio, M., Gong, S., Xiang, T.: Recognizing action as clouds of space-time interest points. In: Conference on Computer Vision and Pattern Recognition (2009)
6. Cao, Y., Barrett, D., Barbu, A., Narayanaswamy, S., Yu, H., Michaux, A., Lin, Y., Dickinson, S., Siskind, J., Wang, S.: Recognizing human activities from partially observed videos. In: Conference on Computer Vision and Pattern Recognition (2013)
7. Chen, C.-C., Aggarwal, J.K.: Recognizing human action from a far field of view. In: Workshop on Motion and Video Computing, 2009, WMVC '09 (2009)
8. Choi, W., Savarese, S.: A unified framework for multi-target tracking and collective activity recognition. In: European Conference on Computer Vision, pp. 215–230. Springer, Berlin (2012)
9. Choi, W., Shahid, K., Savarese, S.: What are they doing?: collective activity classification using spatio-temporal relationship among people. In: International Conference on Computer Vision Workshops, pp. 1282–1289 (2009)
10. Choi, W., Shahid, K., Savarese, S.: Learning context for collective activity recognition. In: Conference on Computer Vision and Pattern Recognition (2011)
11. Dalal, N., Triggs, B.: Histograms of oriented gradients for human detection. In: Conference on Computer Vision and Pattern Recognition (2005)
12. Desai, C., Ramanan, D., Fowlkes, C.: Discriminative models for multi-class object layout. In: International Conference on Computer Vision (2009)
13. Desai, C., Ramanan, D., Fowlkes, C.: Discriminative models for static human-object interactions. In: Conference on Computer Vision and Pattern Recognition Workshop on Structured Models in Computer Vision (2010)
14. Dollar, P., Rabaud, V., Cottrell, G., Belongie, S.: Behavior recognition via sparse spatio-temporal features. In: Visual Surveillance and Performance Evaluation of Tracking and Surveillance (2005)
15. Efros, A.A., Berg, A.C., Mori, G., Malik, J.: Recognizing action at a distance. In: International Conference on Computer Vision, vol. 2, pp. 726–733 (2003)
16. Filipovych, R., Ribeiro, E.: Recognizing primitive interactions by exploring actor-object states. In: Conference on Computer Vision and Pattern Recognition, pp. 1–7. IEEE, New York (2008)
17. Gong, S., Xiang, T.: Recognition of group activities using dynamic probabilistic networks. In: International Conference on Computer Vision, vol. 2, pp. 742–749 (2003)
18. Gorelick, L., Blank, M., Shechtman, E., Irani, M., Basri, R.: Actions as space-time shapes. IEEE Trans. Pattern Anal. Mach. Intell. 29(12), 2247–2253 (2007)
19. Gupta, A., Kembhavi, A., Davis, L.S.: Observing human-object interactions: using spatial and functional compatibility for recognition. IEEE Trans. Pattern Anal. Mach. Intell. 31(10), 1775–1789 (2009)
20. Hadfield, S., Bowden, R.: Hollywood 3d: recognizing actions in 3d natural scenes. In: Conference on Computer Vision and Pattern Recognition, Portland (2013)
21. Hasan, M., Roy-Chowdhury, A.K.: Continuous learning of human activity models using deep nets. In: European Conference on Computer Vision (2014)
22. Hoai, M., De la Torre, F.: Max-margin early event detectors. In: Conference on Computer Vision and Pattern Recognition (2012)
23. Hoai, M., Lan, Z.-Z., De la Torre, F.: Joint segmentation and classification of human actions in video. In: Conference on Computer Vision and Pattern Recognition (2011)
24. Jain, M., Jégou, H., Bouthemy, P.: Better exploiting motion for better action recognition. In: Conference on Computer Vision and Pattern Recognition (2013)
25. Ji, S., Xu, W., Yang, M., Yu, K.: 3d convolutional neural networks for human action recognition. In: International Conference on Machine Learning (2010)

26. Ji, S., Xu, W., Yang, M., Yu, K.: 3d convolutional neural networks for human action recognition. IEEE Trans. Pattern Anal. Mach. Intell. **35**(1), 221–231 (2013)
27. Jia, C., Kong, Y., Ding, Z., Fu, Y.: Latent tensor transfer learning for RGB-D action recognition. In: ACM Multimedia (2014)
28. Liu, J.L.J., Shah, M.: Recognizing realistic actions from videos "in the wild". In: Conference on Computer Vision and Pattern Recognition, vol. 35, pp. 1798–1828 (2009)
29. Karpathy, A., Toderici, G., Shetty, S., Leung, T., Sukthankar, R., Fei-Fei, L.: Large-scale video classification with convolutional neural networks. In: Conference on Computer Vision and Pattern Recognition (2014)
30. Kitani, K.M., Ziebart, B.D., Bagnell, J.A., Hebert, M.: Activity forecasting. In: European Conference on Computer Vision (2012)
31. Klaser, A., Marszalek, M., Schmid, C.: A spatio-temporal descriptor based on 3d-gradients. In: British Machine Vision Conference (2008)
32. Kliper-Gross, O., Hassner, T., Wolf, L.: The action similarity labeling challenge. IEEE Trans. Pattern Anal. Mach. Intell. **34**(3), pp. 615–621 (2012)
33. Kong, Y., Fu, Y.: Modeling supporting regions for close human interaction recognition. In: European Conference on Computer Vision Workshop (2014)
34. Kong, Y., Fu, Y.: Bilinear heterogeneous information machine for RGB-D action recognition. In: Conference on Computer Vision and Pattern Recognition (2015)
35. Kong, Y., Fu, Y.: Max-margin action prediction machine. IEEE Trans. Pattern Anal. Mach. Intell. (2015)
36. Kong, Y., Jia, Y., Fu, Y.: Learning human interaction by interactive phrases. In: European Conference on Computer Vision (2012)
37. Kong, Y., Kit, D., Fu, Y.: A discriminative model with multiple temporal scales for action prediction. In: European Conference on Computer Vision (2014)
38. Kong, Y., Jia, Y., Fu, Y.: Interactive phrases: semantic descriptions for human interaction recognition. IEEE Trans. Pattern Anal. Mach. Intell. vol. 36, pp. 1775–1788 (2014)
39. Koppula, H.S., Gupta, R., Saxena, A.: Learning human activities and object affordances from RGB-D videos. Int. J. Robot. Res. vol. 32, pp. 951–970 (2013)
40. Kovashka, A., Grauman, K.: Learning a hierarchy of discriminative space-time neighborhood features for human action recognition. In: Conference on Computer Vision and Pattern Recognition, pp. 2046–2053. IEEE, New York (2010)
41. Kuehne, H., Jhuang, H., Garrote, E., Poggio, T., Serre, T.: HMDB: A large video database for human motion recognition. In: International Conference on Computer Vision (2011)
42. Kurakin, A., Zhang, Z., Liu, Z.: A real-time system for dynamic hand gesture recognition with a depth sensor. In: European Signal Processing Conference (2012)
43. Lan, T., Wang, Y., Yang, W., Robinovitch, S.N., Mori, G.: Discriminative latent models for recognizing contextual group activities. IEEE Trans. Pattern Anal. Mach. Intell. **34**(8), 1549–1562 (2012)
44. Laptev, I.: On space-time interest points. Int. J. Comput. Vis. **64**(2), 107–123 (2005)
45. Laptev, I., Lindeberg, T.: Space-time interest points. In: International Conference on Computer Vision, pp. 432–439 (2003)
46. Laptev, I., Marszałek, M., Schmid, C., Rozenfeld, B.: Learning realistic human actions from movies. In: Conference on Computer Vision and Pattern Recognition (2008)
47. Le, Q.V., Zou, W.Y., Yeung, S.Y., Ng, A.Y.: Learning hierarchical invariant spatio-temporal features for action recognition with independent subspace analysis. In: Conference on Computer Vision and Pattern Recognition (2011)
48. Li, R., Chellappa, R., Zhou, S.K.: Learning multi-modal densities on discriminative temporal interaction manifold for group activity recognition. In: Conference on Computer Vision and Pattern Recognition, pp. 2450–2457 (2009)
49. Li, W., Zhang, Z., Liu, Z.: Action recognition based on a bag of 3d points. In: Conference on Computer Vision and Pattern Recognition Workshop (2010)
50. Li, K., Hu, J., Fu, Y.: Modeling complex temporal composition of actionlets for activity prediction. In: European Conference on Computer Vision (2012)

51. Li, S., Li, K., Fu, Y.: Temporal subspace clustering for human motion segmentation. In: International Conference on Computer Vision (2015)
52. Liu, L., Shao, L.: Learning discriminative representations from RGB-D video data. In: International Joint Conference on Artificial Intelligence (2013)
53. Liu, J., Ali, S., Shah, M.: Recognizing human actions using multiple features. In: Conference on Computer Vision and Pattern Recognition (2008)
54. Liu, J., Luo, J., Shah, M.: Recognizing realistic actions from videos "in the wild". In: Proceedings of IEEE Conference on Computer Vision and Pattern Recognition (2009)
55. Liu, J., Kuipers, B., Savarese, S.: Recognizing human actions by attributes. In: Conference on Computer Vision and Pattern Recognition (2011)
56. Lucas, B.D., Kanade, T.: An iterative image registration technique with an application to stereo vision. In: International Joint Conference on Artificial Intelligence, pp. 674–679 (1981)
57. Marszałek, M., Laptev, I., Schmid, C.: Actions in context. In: Proceedings of IEEE Conference on Computer Vision and Pattern Recognition (2009)
58. Mehran, R., Oyama, A., Shah, M.: Abnormal crowd behavior detection using social force model. In: Conference on Computer Vision and Pattern Recognition (2009)
59. Ni, B., Yan, S., Kassim, A.A.: Recognizing human group activities with localized causalities. In: Conference on Computer Vision and Pattern Recognition, pp. 1470–1477 (2009)
60. Niebles, J.C., Chen, C.-W., Fei-Fei, L.: Modeling temporal structure of decomposable motion segments for activity classification. In: European Conference on Computer Vision (2010)
61. Odashima, S., Shimosaka, M., Kaneko, T., Fuikui, R., Sato, T.: Collective activity localization with contextual spatial pyramid. In: European Conference on Computer Vision (2012)
62. Oliver, N.M., Rosario, B., Pentland, A.P.: A Bayesian computer vision system for modeling human interactions. IEEE Trans. Pattern Anal. Mach. Intell. **22**(8), 831–843 (2000)
63. Oreifej, O., Liu, Z.: Hon4d: histogram of oriented 4d normals for activity recognition from depth sequences. In: Conference on Computer Vision and Pattern Recognition (2013)
64. Patron-Perez, A., Marszalek, M., Zissermann, A., Reid, I.: High five: recognising human interactions in tv shows. In: British Machine Vision Conference (2010)
65. Patron-Perez, A., Marszalek, M., Reid, I., Zissermann, A.: Structured learning of human interaction in tv shows. IEEE Trans. Pattern Anal. Mach. Intell. **34**(12), 2441–2453 (2012)
66. Pl otz, T., Hammerla, N.Y., Olivier, P.: Feature learning for activity recognition in ubiquitous computing. In: International Joint Conference on Artificial Intelligence (2011)
67. Raptis, M., Sigal, L.: Poselet key-framing: a model for human activity recognition. In: Conference on Computer Vision and Pattern Recognition (2013)
68. Raptis, M., Soatto, S.: Tracklet descriptors for action modeling and video analysis. In: European Conference on Computer Vision (2010)
69. Reddy, K.K., Shah, M.: Recognizing 50 human action categories of web videos. Mach. Vis. Appl. vol. 24, pp 971–981 (2012)
70. Rodriguez, M.D., Ahmed, J., Shah, M.: Action mach: a spatio-temporal maximum average correlation height filter for action recognition. In: Conference on Computer Vision and Pattern Recognition (2008)
71. Ryoo, M.S.: Human activity prediction: early recognition of ongoing activities from streaming videos. In: International Conference on Computer Vision (2011)
72. Ryoo, M.S., Aggarwal, J.K.: Recognition of composite human activities through context-free grammar based representation. In: Conference on Computer Vision and Pattern Recognition, vol. 2, pp. 1709–1718 (2006)
73. Ryoo, M.S., Aggarwal, J.K.: Spatio-temporal relationship match: video structure comparison for recognition of complex human activities. In: International Conference on Computer Vision, pp. 1593–1600 (2009)
74. Ryoo, M., Aggarwal, J.: Ut-interaction dataset, ICPR contest on semantic description of human activities. http://cvrc.ece.utexas.edu/SDHA2010/Human_Interaction.html (2010)
75. Schuldt, C., Laptev, I., Caputo, B.: Recognizing human actions: a local SVM approach. In: Indian Council of Philosophical Research, vol. 3, pp. 32–36. IEEE, New York (2004)

76. Scovanner, P., Ali, S., Shah, M.: A 3-dimensional sift descriptor and its application to action recognition. In: Proceedings of ACM Multimedia (2007)
77. Shechtman, E., Irani, M.: Space-time behavior based correlation. In: Conference on Computer Vision and Pattern Recognition, vol. 1, pp. 405–412. IEEE, New York (2005)
78. Shi, Q., Cheng, L., Wang, L., Smola, A.: Human action segmentation and recognition using discriminative semi-Markov models. Int. J. Comput. Vis. **93**, 22–32 (2011)
79. Simonyan, K., Zisserman, A.: Two-stream convolutional networks for action recognition in videos. In: Conference on Neural Information Processing Systems, (2014)
80. Singh, S.A.V.S., Ragheb, H.: Muhavi: a multicamera human action video dataset for the evaluation of action recognition methods. In: 2nd Workshop on Activity Monitoring by Multi-Camera Surveillance Systems (AMMCSS), pp. 48–55 (2010)
81. Soomro, K., Zamir, A.R., Shah, M.: Ucf101: a dataset of 101 human action classes from videos in the wild (2012). CRCV-TR-12-01
82. Sun, J., Wu, X., Yan, S., Cheong, L.F., Chua, T.S., Li, J.: Hierarchical spatio-temporal context modeling for action recognition. In: Conference on Computer Vision and Pattern Recognition (2009)
83. Sun, L., Jia, K., Chan, T.-H., Fang, Y., Wang, G., Yan, S.: DL-SFA: deeply-learned slow feature analysis for action recognition. In: Conference on Computer Vision and Pattern Recognition (2014)
84. Sung, J., Ponce, C., Selman, B., Saxena, A.: Human activity detection from RGBD images. In: AAAI Workshop on Pattern, Activity and Intent Recognition (2011)
85. Tang, K., Fei-Fei, L., Koller, D.: Learning latent temporal structure for complex event detection. In: Conference on Computer Vision and Pattern Recognition (2012)
86. Taylor, G.W., Fergus, R., LeCun, Y., Bregler, C.: Convolutional learning of spatio-temporal features. In: European Conference on Computer Vision (2010)
87. Vahdat, A., Gao, B., Ranjbar, M., Mori, G.: A discriminative key pose sequence model for recognizing human interactions. In: International Conference on Computer Vision Workshops, pp. 1729–1736 (2011)
88. Wang, Y., Mori, G.: Hidden part models for human action recognition: probabilistic vs. max-margin. IEEE Trans. Pattern Anal. Mach. Intell. vol. 33, pp. 1310–1323 (2010)
89. Wang, H., Schmid, C.: Action recognition with improved trajectories. In: IEEE International Conference on Computer Vision, Sydney (2013). http://hal.inria.fr/hal-00873267
90. Wang, H., Ullah, M.M., Klaser, A., Laptev, I., Schmid, C.: Evaluation of local spatio-temporal features for action recognition. In: British Machine Vision Conference (2008)
91. Wang, H., Ullah, M.M., Klaser, A., Laptev, I., Schmid, C.: Evaluation of local spatio-temporal features for action recognition. In: British Machine Vision Conference (2009)
92. Wang, H., Kläser, A., Schmid, C., Liu, C.-L.: Action recognition by dense trajectories. In: IEEE Conference on Computer Vision & Pattern Recognition, pp. 3169–3176, Colorado Springs (2011). http://hal.inria.fr/inria-00583818/en
93. Wang, J., Liu, Z., Wu, Y., Yuan, J.: Mining actionlet ensemble for action recognition with depth cameras. In: Conference on Computer Vision and Pattern Recognition (2012)
94. Wang, J., Liu, Z., Chorowski, J., Chen, Z., Wu, Y.: Robust 3d action recognition with random occupancy patterns. In: European Conference on Computer Vision (2012)
95. Wang, Z., Wang, J., Xiao, J., Lin, K.-H., Huang, T.S.: Substructural and boundary modeling for continuous action recognition. In: Conference on Computer Vision and Pattern Recognition (2012)
96. Wang, H., Kla aser, A., Schmid, C., Liu, C.-L.: Dense trajectories and motion boundary descriptors for action recognition. Int. J. Comput. Vis. **103** pp. 60–79 (2013)
97. Wang, K., Wang, X., Lin, L., Wang, M., Zuo, W.: 3d human activity recognition with reconfigurable convolutional neural networks. In: ACM Multimedia (2014)
98. Wang, H., Oneata, D., Verbeek, J., Schmid, C.: A robust and efficient video representation for action recognition. Int. J. Comput. Vis. (2015)
99. Wang, P., Li, W., Gao, Z., Tang, J.Z.C., Ogunbona, P.: Deep convolutional neural networks for action recognition using depth map sequences. arXiv:1501.04686

100. Weinland, D., Ronfard, R., Boyer, E.: Free viewpoint action recognition using motion history volumes. Comput. Vis. Image Underst. **104**(2–3), pp. 249–257 (2006)
101. Willems, G., Tuytelaars, T., Gool, L.: An efficient dense and scale-invariant spatio-temporal interest point detector. In: European Conference on Computer Vision (2008)
102. Xia, L., Aggarwal, J.K.: Spatio-temporal depth cuboid similarity feature for activity recognition using depth camera. In: Conference on Computer Vision and Pattern Recognition (2013)
103. Xia, L., Chen, C., Aggarwal, J.K.: View invariant human action recognition using histograms of 3d joints. In: 2012 IEEE Computer Society Conference on Computer Vision and Pattern Recognition Workshops (CVPRW), pp. 20–27. IEEE, New York (2012)
104. Yang, Y., Shah, M.: Complex events detection using data-driven concepts. In: European Conference on Computer Vision (2012)
105. Yao, B., Fei-Fei, L.: Modeling mutual context of object and human pose in human-object interaction activities. In: Conference on Computer Vision and Pattern Recognition, pp. 17–24 (2010)
106. Yao, B., Fei-Fei, L.: Action recognition with exemplar based 2.5d graph matching. In: European Conference on Computer Vision (2012)
107. Yao, B., Fei-Fei, L.: Recognizing human-object interactions in still images by modeling the mutual context of objects and human poses. IEEE Trans. Pattern Anal. Mach. Intell. **34**(9), 1691–1703 (2012)
108. Yeffet, L., Wolf, L.: Local trinary patterns for human action recognition. In: Conference on Computer Vision and Pattern Recognition (2009)
109. Yu, T.-H., Kim, T.-K., Cipolla, R.: Real-time action recognition by spatiotemporal semantic and structural forests. In: British Machine Vision Conference (2010)
110. Yu, G., Liu, Z., Yuan, J.: Discriminative orderlet mining for real-time recognition of human-object interaction. In: Asian Conference on Computer Vision (2014)
111. Yuan, J., Liu, Z., Wu, Y.: Discriminative subvolume search for efficient action detection. In: Conference on Computer Vision and Pattern Recognition (2009)
112. Yuan, J., Liu, Z., Wu, Y.: Discriminative video pattern search for efficient action detection. IEEE Trans. Pattern Anal. Mach. Intell. (2010)

Chapter 2
Action Recognition and Human Interaction

Yu Kong and Yun Fu

1 Introduction

Recognizing human activities is a fundamental problem in the computer vision community and is a key step toward the automatic understanding of scenes. Compared with single-person action [28, 32, 48], human interaction is a typical human activity in real world and has received much attention in the community [2, 24, 35, 41].

As previous work [3, 6, 16, 24, 50] shows, integrating co-occurrence information from various perceptual tasks such as object recognition, object location, and human pose estimation is helpful for disambiguating visually similar interactions and facilitating the recognition task. However, misclassifications remain in some challenging situations. This suggests that the co-occurrence relationships are not expressive enough to deal with interactions containing large variations. For example, in "boxing" interaction, the defender can perform diverse semantic actions to protect himself, e.g. step back, crouch, or even hit back. This requires us to define all possible action co-occurrence relationships and provide sufficient training data for each co-occurrence case, which is computationally expensive.

In addition to that, interaction videos often contain close interactions between multiple people with physical contact (e.g., "hug" and "fight"). This raises two major challenges in understanding this type of interaction videos: the body part

Y. Kong (✉)
Department of Electrical and Computer Engineering, Northeastern University, 360 Huntington Avenue, Boston, MA 02115, USA
e-mail: yukong@ece.neu.edu

Y. Fu
Department of Electrical and Computer Engineering and College of Computer and Information Science (Affiliated), Northeastern University, 360 Huntington Avenue, Boston, MA 02115, USA
e-mail: yunfu@ece.neu.edu

© Springer International Publishing Switzerland 2016
Y. Fu (ed.), *Human Activity Recognition and Prediction*,
DOI 10.1007/978-3-319-27004-3_2

occlusion and the ambiguity in feature assignments (features such as interest points are difficult to be uniquely assigned to a particular person in close interactions). Unfortunately, the aforementioned problems are not addressed in existing interaction recognition methods [2, 24, 25, 51]. Methods in [2, 43] use trackers/detectors to roughly extract people, and assume interactions do not contain close physical contact (e.g., "walk" and "talk"). Their performance is limited in close interactions since the feature of one single person may contain noises from background or the other interacting people. Feature assignment problem is avoided in [25, 51] by treating the interaction people as a group. However, they do not utilize the intrinsic rich context of the interaction. Interest points have shown that they can be mainly associated with foreground moving human bodies in conventional single-person action recognition methods [28]. However, since multiple people present in interactions, it is difficult to accurately assign interest points to a single person, especially in close interactions. Therefore, action representations of people are extremely noisy and consequently degrade the recognition performance.

In this chapter, we propose two approaches to address the two challenges mentioned above. The first approach uses semantic descriptions, *interactive phrases*, to better describe complex interactions. The second approach learns discriminative spatiotemporal patches to better separate the interacting people, in order to provide cleaner features for interaction recognition.

2 Approach I: Learning Semantic Descriptions for Interaction Recognition

As suggested in [28], human action categories share a set of basic action components or action primitives. These action primitives allow us to represent a variety of human actions. Motivated by this idea, we present *interactive phrases* [18, 19] to build primitives for representing human interactions. Essentially, these phrases[1] are descriptive primitives shared by all interaction classes. They describe binary motion relationships and characterize an interaction from different angles, e.g. motion relationships between arms, legs, torsos, etc. (Fig. 2.1). Consequently, we can simply use compositions of phrases to describe interactions with variations rather than considering all possible action co-occurrences in an interaction class.

The significance of interactive phrases is that they incorporate rich human knowledge about motion relationships. The use of the knowledge allows us to better represent complex human interactions and elegantly integrate the knowledge into discriminative models. Moreover, we treat phrases as mid-level features to bridge the gap between low-level features and high-level interaction classes, which can improve recognition performance. In addition, phrases provide a novel type of contextual information, i.e. phrase context, for human interaction recognition. Since

---

[1]In this chapter, we use interactive phrases and phrases interchangeably.

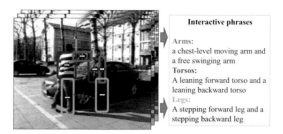

Interactive phrases

Arms:
a chest-level moving arm and
a free swinging arm
Torsos:
A leaning forward torso and a
leaning backward torso
Legs:
A stepping forward leg and a
stepping backward leg

Fig. 2.1 The learned phrases from a video. Phrases express semantic motion relationships between interacting people. For example, the motion relationships between people's arms (in *blue boxes*), people's torsos (in *purple boxes*), or people's legs (in *green boxes*). The *red dotted lines* indicate that there are some relationships between the two parts and *yellow arrows* show the directions of motion

phrases describe all the important details of an interaction, they provide a strong context for each other and are more expressive than the action context used in [24]. Interactive phrases allow us to build a more descriptive model, which can be used to recognize human interactions with large variations (e.g., interactions with partial occlusion).

Interactive phrases and attributes of objects [11, 22, 47] and actions [28] share some similarities. They are all descriptive primitives and introduce high-level semantic knowledge into predictive models. The main difference is that attributes represent *unary* relationships of an object while phrases describe high-order relationships between people. In this work, we focus on *binary* relationships in human interactions. In other words, attributes of objects focus on the intrinsic properties of an object (e.g. "furry," "metal"), while phrases provide an effective way to describe motion relationships between interacting people. Phrases can also be regarded as activity primitives which allow us to recognize novel interaction classes. The proposed phrases are philosophically similar to the bag-of-words model in which several words can be treated as a primitive to represent an action. However, our phrases naturally incorporate rich human knowledge into our model to better represent human interactions. Moreover, phrases describe motion relationships between people, and are used as mid-level features, which bridge the semantic gap between low-level features and high-level interactions.

2.1 Related Work

A considerable amount of previous work in human action recognition focuses on recognizing the actions of a single person in videos [1, 14, 20, 32, 43]. However, in most of the previous studies [25, 27, 29, 39, 51], interactions are recognized in the same way as single-person actions. Specifically, interactions are represented as a motion descriptor including all the people in a video, and then an action classifier is adopted to classify interactions. Although reasonable performance can be achieved, they do not utilize the intrinsic properties of interactions, such as co-occurrence information between interacting people. In addition, they treat people as a single entity and do not extract the motion of each person from the group. Thus, their method could not give the action label of each interacting person in the video simultaneously.

Exploiting rich contextual information in human interactions can help achieve more accurate and robust results. In [34], human interactions are recognized using motion co-occurrence, which is achieved by coupling motion state of one person with the other interacting person. Spatiotemporal crowd context is captured in [3] to recognize human interactions. These two studies are limited to the coherent behavior of individuals in time and space. In [35], spatial relationships between individuals is captured using the structured learning technique. Individuals in interactions can also be represented by a set of key poses, and their spatial and temporal relationships can be captured for recognition [45]. Co-occurrence between interaction and atomic action, and interaction and collective action is captured for interaction recognition [2]. In these studies, the contextual information is easy to capture since the human body size in their datasets is not very small and thus a visual tracker for human body is performed well. However, in [26, 30, 31], the size of human body can often be very small, making them difficult or even impossible to track using tracker. Consequently, the contextual information between interacting entities is difficult to extract.

Contextual information has also been effectively exploited in human–object interactions and object–object interactions. Human–object interactions in videos and images are performed in [16] by fusing context from object reaction, object class, and manipulation motion into a model. Mutual context of objects and human poses is explored in [49] for human–object interaction recognition. In [13], temporal and casual relationships between object events are represented by the interconnected multiple hidden Markov models. A contextual model is proposed to capture the relative locations of objects and human poses [6]. The relationships between human and object are modeled by joint actor–object states [12]. A discriminative model is proposed to capture the spatial relationships (e.g., above, below) between objects for multi-object detection [5].

However, to the best of our knowledge, few attempts have been made to utilize high-level descriptions for human interaction recognition. A related work to ours is [41] in which the context-free grammar is applied to describe spatiotemporal relationships between people. The key difference between our work and theirs is that our method integrates high-level descriptions and interaction classes into a unified model. In addition, these descriptions (interactive phrases) are treated as latent variables to deal with intra-class variability. Our work is also different from [24]. Our model depends on high-level descriptions (interactive phrases), while [24] relies on action co-occurrence. Our work decomposes action co-occurrence into phrase co-occurrence, which provides a more effective way to represent complex interactions.

The descriptive power of high-level description-based methods has been demonstrated in object recognition [9, 22, 47], object detection [42], and action recognition [28]. These approaches utilize attributes to describe intrinsic properties of an object (e.g. color, shape) or spatial–temporal visual characteristics of an actor (e.g., single leg motion). Our interactive phrases are different from attributes of objects [47] and actions [28]. In their work, attributes represent unary relationships (intrinsic properties of an object or an action), which are directly inferred from low-level features. By contrast, interactive phrases describe binary motion relationships and are built based on semantic motion attributes of each interacting person.

Our work is partially inspired by Gupta and Davis [15] which used language constructs such as "prepositions" (e.g. above, below) and "comparative adjectives" (e.g., brighter, smaller) to express relationships between objects. The difference is that our interactive phrases describe motion relationships of people rather than spatial relationships of static objects. Moreover, interactive phrases are built upon semantic motion attributes rather than being inferred from object classes.

2.2 Our Models

Our method consists of two main components, the attribute model and the interaction model (Fig. 2.2). The attribute model is utilized to jointly detect all attributes for each person, and the interaction model is applied to recognize an interaction. In this work, we mainly focus on recognizing interactions between two people.

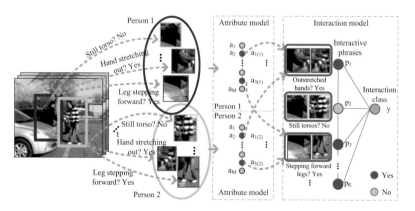

Fig. 2.2 Framework of our interactive phrase method

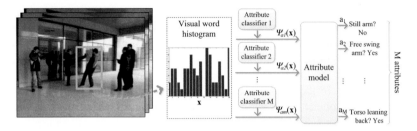

Fig. 2.3 Framework of detecting motion attributes from videos

2.2.1 Attribute Model

We utilize motion attributes to describe individual actions [28], e.g. "arm raising up motion," "leg stepping backward motion," etc. In interactions, both of the two interacting people have the same attribute vocabulary but with different values. Those motion attributes can be inferred from low-level motion features (Fig. 2.3), for example, spatiotemporal interest points [8]. We assume there are certain interdependencies between attribute pairs (a_j, a_k). For instance, attributes "arm stretching out motion" and "leg stepping forward motion" tend to appear together in "handshake." The interdependencies are greatly helpful in dealing with incorrect attributes caused by motion ambiguity and partial occlusion.

We adopt a tree-structured undirected graph [21] $\mathcal{G}_a = (\mathcal{V}_a, \mathcal{E}_a)$ to represent the configurations of attributes. A vertex $a_m \in \mathcal{V}_a$ ($m = 1, \ldots, M$) corresponds to the mth attribute and an edge $(a_j, a_k) \in \mathcal{E}_a$ corresponds to the dependency between the two attributes. We use a discriminative function $g_\lambda : \mathbb{X} \times \mathbb{A} \to \mathbb{R}$ to score each training example $(\mathbf{x}, \mathbf{a})$: $g_\lambda(\mathbf{x}, \mathbf{a}) = \boldsymbol{\lambda}^\mathrm{T} \Phi(\mathbf{x}, \mathbf{a})$, where $\mathbf{x}$ denotes the feature of a person in an interaction and $\mathbf{a} = (a_1, \ldots, a_M)$ is a binary attribute vector. $a_m = 0$ means the mth attribute is absent and $a_m = 1$ denotes the attribute is present. We define $\boldsymbol{\lambda}^\mathrm{T} \Phi(\mathbf{x}, \mathbf{a})$ as a summation of potential functions:

$$\boldsymbol{\lambda}^\mathrm{T} \Phi(\mathbf{x}, \mathbf{a}) = \sum_{a_j \in \mathcal{V}_a} \boldsymbol{\lambda}_{a_j}^\mathrm{T} \phi_1(\mathbf{x}, a_j) + \sum_{(a_j, a_k) \in \mathcal{E}_a} \boldsymbol{\lambda}_{a_j a_k}^\mathrm{T} \phi_2(a_j, a_k), \qquad (2.1)$$

where $\boldsymbol{\lambda} = \{\boldsymbol{\lambda}_{a_j}, \boldsymbol{\lambda}_{a_j a_k}\}$ is model parameter. In our work, graph structure $\mathcal{E}_a$ is learned by the Chow–Liu algorithm [4]. The potential functions in Eq. (2.1) are summarized as follows.

Unary potential $\boldsymbol{\lambda}_{a_j}^\mathrm{T} \phi_1(\mathbf{x}, a_j)$ provides the score for an attribute a_j and is used to indicate the presence of a_j given the motion feature $\mathbf{x}$. Parameter $\boldsymbol{\lambda}_{a_j}$ is a template for an attribute a_j. The feature function $\phi_1(\mathbf{x}, a_j)$ models the agreement between observation $\mathbf{x}$ and motion attribute a_j, and is given by

$$\phi_1(\mathbf{x}, a_j) = \delta(a_j = u) \cdot \psi_{a_j}(\mathbf{x}). \qquad (2.2)$$

Here, $\delta(\cdot)$ denotes an indicator function, $u \in \mathcal{A}$ denotes a state of the attribute a_j, where $\mathcal{A}$ is the attribute space. Instead of keeping $\psi_{a_j}(\mathbf{x})$ as a high-dimensional feature vector, we represent it as the score output of a linear SVM trained with attribute a_j. Similar tricks have been used in [5, 47].

Pairwise potential $\boldsymbol{\lambda}_{a_j a_k}^{\mathrm{T}} \phi_2(a_j, a_k)$ captures the co-occurrence of a pair of attributes a_j and a_k, for example, the co-occurrence relationships between attributes "torso bending motion" and "still leg" in "bow." Parameter $\boldsymbol{\lambda}_{a_j a_k}$ is a 4-dimensional vector representing the weights for all configurations of a pair of attributes. The feature function $\phi_2(a_j, a_k)$ models the co-occurrence relationships of two attributes. We define $\phi_2(a_j, a_k)$ for a co-occurrence (u, v) as

$$\phi_2(a_j, a_k) = \delta(a_j = u) \cdot \delta(a_k = v). \tag{2.3}$$

2.2.2 Interaction Model

Interactive phrases encode human knowledge about motion relationships between people. The phrases are built on attributes of two interacting people and utilized to describe their co-occurrence relationships. Let p_j be the jth phrase associated with two people's attributes $a_{j(1)}$ and $a_{j(2)}$.[2] In the interaction model, we use $a_{j(i)}$ to denote the attribute of the ith person that links to the jth phrase. For example, phrase p_j "cooperative interaction" is associated with two people's attributes $a_{j(1)}$ and $a_{j(2)}$ "friendly motion." Note that $a_{j(i)}$ and $a_{k(i)}$ could be the same attribute but link to different phrases. We also assume that there are certain interdependencies between some phrase pairs (p_j, p_k). For example, phrases "interaction between stretching out hands" and "interaction between stepping forward legs" are highly correlated in "handshake" (Fig. 2.4).

We employ an undirected graph $\mathcal{G}_p = (\mathcal{V}_p, \mathcal{E}_p)$ to encode the configurations of phrases. A vertex $p_j \in \mathcal{V}_p$ $(j = 1, \dots, K)$ corresponds to the jth phrase and an edge $(p_j, p_k) \in \mathcal{E}_p$ corresponds to the dependency between the two phrases. Note that intra-class variability leads to different interactive phrase values in certain interaction classes. For instance, in "handshake," some examples have interactive phrase p_j "interaction between stepping forward legs" but some do not. In addition, labeling attributes is a subjective process and thus would influence the values of interactive phrases. We deal with this problem by treating phrases as latent variables and formulating the classification problem based on the latent SVM framework [10, 46].

Given training examples $\{\hat{\mathbf{x}}^{(n)}, y^{(n)}\}_{n=1}^N$, we are interested in learning a discriminative function $f_{\mathbf{w}}(\hat{\mathbf{x}}, \hat{\mathbf{a}}, y) = \max_{\mathbf{p}} \mathbf{w}^{\mathrm{T}} \Phi(\hat{\mathbf{x}}, \hat{\mathbf{a}}, \mathbf{p}, y)$. Here $\hat{\mathbf{x}} = (\mathbf{x}_1, \mathbf{x}_2)$ is raw features of two interacting people, $\hat{\mathbf{a}} = (\mathbf{a}_1, \mathbf{a}_2)$ denotes two people's attributes, $\mathbf{p} = (p_1, \dots, p_K)$ is a binary vector of phrases, and y is an interaction class,

---

[2]Please refer to the supplemental material to see details about the connectivity patterns of interactive phrases and attributes.

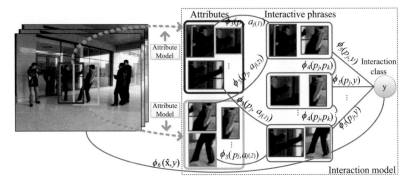

Fig. 2.4 The unary, pairwise, and global interaction potentials in the interaction model

where $p_k \in \mathcal{P}$ and $y \in \mathcal{Y}$. To obtain $\mathbf{a}_1$ and $\mathbf{a}_2$, we run the attribute model twice with corresponding features. We define $\mathbf{w}^T \Phi(\hat{\mathbf{x}}, \hat{\mathbf{a}}, \mathbf{p}, y)$ as a summation of potential functions: (see Fig. 2.4)

$$\mathbf{w}^T \Phi(\hat{\mathbf{x}}, \hat{\mathbf{a}}, \mathbf{p}, y) = \sum_{p_j \in \mathcal{V}_p} \sum_{i=1}^{2} \mathbf{w}^T_{p_j a_{j(i)}} \phi_3(p_j, a_{j(i)}) + \sum_{p_j \in \mathcal{V}_p} \mathbf{w}^T_{p_j y} \phi_4(p_j, y)$$

$$+ \sum_{(p_j, p_k) \in \mathcal{E}_p} \mathbf{w}^T_{p_j p_k} \phi_5(p_j, p_k) + \mathbf{w}^T_{\hat{\mathbf{x}} y} \phi_6(\hat{\mathbf{x}}, y), \qquad (2.4)$$

where $\mathbf{w} = \{\mathbf{w}_{p_j a_{j(i)}}, \mathbf{w}_{p_j p_k}, \mathbf{w}_{p_j y}, \mathbf{w}_{\hat{\mathbf{x}}}\}$ is model parameter. Similar to the attribute model, we use the Chow–Liu algorithm [4] to learn graph structure $\mathcal{E}_p$ in the interaction model. The potential functions are enumerated as follows.

Unary potential $\mathbf{w}^T_{p_j a_{j(i)}} \phi_3(p_j, a_{j(i)})$ models the semantic relationships between an interactive phrase p_j and its associated attribute $a_{j(i)}$. Each interactive phrase in this chapter is associated with one attribute of each interacting person. Parameter $\mathbf{w}_{p_j a_{j(i)}}$ is a 4-dimensional vector encoding the weights for all configurations between a phrase and an attribute, and feature function $\phi_3(p_j, a_{j(i)})$ models the agreement between them. The feature function $\phi_3(p_j, a_{j(i)})$ for a configuration (h, u), where $h \in \mathcal{P}$ and $u \in \mathcal{A}$, is given by

$$\phi_3(p_j, a_{j(i)}) = \delta(p_j = h) \cdot \delta(a_{j(i)} = u). \qquad (2.5)$$

Unary potential $\mathbf{w}^T_{p_j y} \phi_4(p_j, y)$ indicates that how likely the interaction class is y and the jth interactive phrase is p_j. Feature function $\phi_4(p_j, y)$ is used to encode the semantic relationships between an interaction class y and a phrase p_j. We define the feature function for a relationship (h, b), where $b \in \mathcal{Y}$, as

$$\phi_4(p_j, y) = \delta(p_j = h) \cdot \delta(y = b). \qquad (2.6)$$

Parameter $\mathbf{w}_{p_j y}$ indicates the weight encoding valid relationships between a phrase p_j and an interaction class y.

Pairwise potential $\mathbf{w}_{p_j p_k}^{\mathrm{T}} \phi_5(p_j, p_k)$ captures the co-occurrence of a pair of interactive phrases (p_j, p_k). Parameter $\mathbf{w}_{p_j p_k}$ is a 4-dimensional vector denoting the weights of all possible configurations of a pair of phrases. Feature function $\phi_5(p_j, p_k)$ in the pairwise potential captures the co-occurrence relationships between two phrases. We define $\phi_5(p_j, p_k)$ for a relationship (h_1, h_2) as

$$\phi_5(p_j, p_k) = \delta(p_j = h_1) \cdot \delta(p_k = h_2). \tag{2.7}$$

Global interaction potential $\mathbf{w}_{\hat{\mathbf{x}}y}^{\mathrm{T}} \phi_6(\hat{\mathbf{x}}, y)$ provides the score measuring how well the raw feature $\hat{\mathbf{x}}$ matches the interaction class template $\mathbf{w}_{\hat{\mathbf{x}}y}$. The feature function $\phi_6(\hat{\mathbf{x}}, y)$ models the dependence between an interaction class with its corresponding video evidence. The feature function for interaction class $y = b$ is defined as

$$\phi_6(\hat{\mathbf{x}}, y) = \delta(y = b) \cdot \hat{\mathbf{x}}. \tag{2.8}$$

2.3 Learning and Inference

Parameter learning in our work consists of two steps: learning parameters of the attribute model and learning parameters of the interaction model.

The max-margin conditional random field formulation [44] is adopted to train the attribute model given training examples $\mathcal{D}_a = \{\mathbf{x}^{(n)}, \mathbf{a}^{(n)}\}_{n=1}^{N_a}$:

$$\min_{\boldsymbol{\lambda}, \xi} \frac{1}{2} \|\boldsymbol{\lambda}\|^2 + C \sum_n \xi_n$$

$$\text{s.t. } \boldsymbol{\lambda}^{\mathrm{T}} \Phi(\mathbf{x}^{(n)}, \mathbf{a}^{(n)}) - \boldsymbol{\lambda}^{\mathrm{T}} \Phi(\mathbf{x}^{(n)}, \mathbf{a}) \geqslant \Delta(\mathbf{a}, \mathbf{a}^{(n)}) - \xi_n, \forall n, \forall \mathbf{a}, \tag{2.9}$$

where C is the trade-off parameter similar to that in SVMs, ξ_n is the slack variable for the nth training example to handle the case of soft margin, and $\Delta(\mathbf{a}, \mathbf{a}^{(n)})$ is the 0-1 loss function.

Next, the latent SVM formulation [10, 46] is employed to train the parameter $\mathbf{w}$ of the interaction model given training examples $\mathcal{D} = \{\hat{\mathbf{x}}^{(n)}, \hat{\mathbf{a}}^{(n)}, y^{(n)}\}_{n=1}^{N}$, where $\hat{\mathbf{a}}^{(n)} = (\mathbf{a}_1^{(n)}, \mathbf{a}_2^{(n)})$ is the attributes of interacting people inferred by the trained attribute model:

$$\min_{\mathbf{w}, \xi} \frac{1}{2} \|\mathbf{w}\|^2 + C \sum_n \xi_n$$

$$\text{s.t. } \max_{\mathbf{p}} \mathbf{w}^{\mathrm{T}} \Phi(\hat{\mathbf{x}}^{(n)}, \hat{\mathbf{a}}^{(n)}, \mathbf{p}, y^{(n)})$$

$$- \max_{\mathbf{p}} \mathbf{w}^{\mathrm{T}} \Phi(\hat{\mathbf{x}}^{(n)}, \hat{\mathbf{a}}^{(n)}, \mathbf{p}, y) \geqslant \Delta(y, y^{(n)}) - \xi_n, \forall n, \forall y. \tag{2.10}$$

This optimization problem can be solved by the coordinate descent [10]. We first randomly initialize the model parameter $\mathbf{w}$ and then learn the parameter $\mathbf{w}$ by iterating the following two steps:

1. Holding $\mathbf{w}$ fixed, find the best interactive phrase configuration $\mathbf{p}'$ such that $\mathbf{w}^T \Phi(\hat{\mathbf{x}}^{(n)}, \mathbf{a}^{(n)}, \mathbf{p}, y^{(n)})$ is maximized.
2. Holding $\mathbf{p}$ fixed, optimize parameter $\mathbf{w}$ by solving the problem [Eq. (2.10)].

In testing, our aim is to infer the interaction class of an unknown example: $y^* = \arg\max_{y \in \mathcal{Y}} f_{\mathbf{w}}(\hat{\mathbf{x}}, \hat{\mathbf{a}}, y)$. However, the attributes $\hat{\mathbf{a}}$ of two interacting people are unknown during testing. We solve this problem by finding the best attribute configuration $\mathbf{a}_i$ for the ith person by running Belief Propagation (BP) in the attribute model: $\mathbf{a}_i = \arg\max_{\mathbf{a}_i} \lambda^T \Phi(\mathbf{x}_i, \mathbf{a}_i)$. Then attributes $\hat{\mathbf{a}} = (\mathbf{a}_1, \mathbf{a}_2)$ is derived and utilized as the input for inferring the interaction class y. BP is also applied to find the best interactive phrase configuration $\hat{\mathbf{p}}$ in the interaction model: $f_{\mathbf{w}}(\hat{\mathbf{x}}, \hat{\mathbf{a}}, y) = \max_{\mathbf{p}} \mathbf{w}^T \Phi(\hat{\mathbf{x}}, \hat{\mathbf{a}}, \mathbf{p}, y)$.

2.4 Experiments

2.4.1 Spatial–Temporal Features

The spatial–temporal interest points [8] are detected from videos of human interaction. The spatial–temporal volumes around the detected points are extracted and represented by gradient descriptors. The dimensionality of gradient descriptors is reduced by PCA. All descriptors are quantized to 1000 visual-words using the k-means algorithm. Then videos are represented by histograms of visual-words.

2.4.2 Dataset

We compile a new dataset, BIT-Interaction dataset, to evaluate our method (see Fig. 2.5) and add a list of 23 interactive phrases based on 17 attributes for all the videos. (Please refer to the supplemental material for details.) Videos are captured in realistic scenes with cluttered background and bounding boxes of interacting people are annotated. People in each interaction class behave totally different and thus have diverse motion attributes (e.g., in some "boxing" videos, people step forward but in some videos they do not). This dataset consists of eight classes of human

Fig. 2.5 Example frames of BIT-interaction dataset. This dataset consists of eight classes of human interactions: bow, boxing, handshake, high-five, hug, kick, pat, and push

interactions (bow, boxing, handshake, high-five, hug, kick, pat, and push), with 50 videos per class. The dataset contains a varied set of challenges including variations in subject appearance, scale, illumination condition, and viewpoint. In addition, in most of the videos, actors are partially occluded by body parts of the other person, poles, bridges, pedestrians, etc. Moreover, in some videos, interacting people have overlapping motion patterns with some irrelevant moving objects in the background (e.g., cars, pedestrians). We randomly choose 272 videos to train the interaction model and use the remaining 128 videos for testing. One hundred and forty-four videos in the training data are utilized to train the attribute model.

2.4.3   Results

We conduct three experiments to evaluate our method. First, we test the proposed method on the BIT-Interaction dataset and compare our method with action context based method [23]. Next, we evaluate the effectiveness of components in the proposed method.

In the first experiment, we test the proposed method on BIT-Interaction dataset. The confusion matrix is shown in Fig. 2.6a. Our method achieves 85.16 % accuracy in classifying human interactions. Some of the classification examples are displayed in Fig. 2.6b. Our method can recognize human interactions in some challenging situations, e.g. partially occlusion and background clutter. This is mainly due to the effect of interdependencies between interactive phrases. In such challenging scenarios, the interdependencies provide a strong context for the incorrectly inferred phrases and thus make them better fit in the context. As a result, human interaction in challenging situations can be correctly recognized. As we show in the last row in Fig. 2.6b, most of the misclassifications are due to visually similar movements in different interaction classes (e.g., "boxing" and "pat") and significant occlusion.

To further investigate the effect of the interdependencies between interactive phrases, we remove the interdependencies $\phi_4(p_j, p_k)$ from the full model and

Fig. 2.6 Results of our method on BIT-interaction dataset. In (**b**), correctly recognized examples are in the *first two rows* and misclassifications are in the *last row*. (**a**) Confusion matrix. (**b**) Classification examples of our method

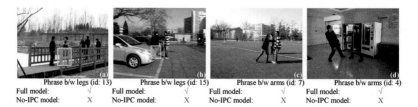

Phrase b/w legs (id: 13)	Phrase b/w legs (id: 15)	Phrase b/w arms (id: 7)	Phrase b/w arms (id: 4)
Full model: √	Full model: √	Full model: √	Full model: √
No-IPC model: X	No-IPC model: X	No-IPC model: X	No-IPC model: X

Fig. 2.7 Classification examples in BIT-interaction dataset with occlusion and background noise. *Yellow boxes* denote occlusion and *red boxes* represent background noise. Please refer to supplemental material for the meaning of phrases according to their id (**a**) Phrase b/w legs (id: 13) (**b**) Phrase b/w legs (id: 15) (**c**) Phrase b/w arms (id: 7) (**d**) Phrase b/w arms (id: 4)

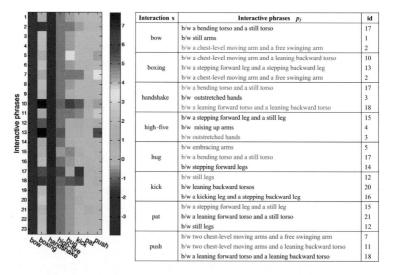

Interaction s	Interactive phrases p_j	id
bow	b/w a bending torso and a still torso	17
	b/w still arms	1
	b/w a chest-level moving arm and a free swinging arm	2
boxing	b/w a chest-level moving arm and a leaning backward torso	10
	b/w a stepping forward leg and a stepping backward leg	13
	b/w a chest-level moving arm and a free swinging arm	2
handshake	b/w a bending torso and a still torso	17
	b/w outstretched hands	3
	b/w a leaning forward torso and a leaning backward torso	18
high-five	b/w a stepping forward leg and a still leg	15
	b/w raising up arms	4
	b/w outstretched hands	3
hug	b/w embracing arms	5
	b/w a bending torso and a still torso	17
	b/w stepping forward legs	14
kick	b/w still legs	12
	b/w leaning backward torsos	20
	b/w a kicking leg and a stepping backward leg	16
pat	b/w a stepping forward leg and a still leg	15
	b/w a leaning forward torso and a still torso	21
	b/w still legs	12
push	b/w two chest-level moving arms and a free swinging arm	7
	b/w two chest-level moving arms and a leaning backward torso	11
	b/w a leaning forward torso and a leaning backward torso	18

Fig. 2.8 (*Left*) The learned importance of different interactive phrases in eight interaction classes. (*Right*) The learned top three important interactive phrases for eight interaction classes, where phrases of significant importance (their weights are at least 10 times greater than the others) are in *blue* and phrases never showed in the training data of an interaction class are in *red*. "b/w" is short for the word "between"

compare the no-IPC model [the full model without $\phi_4(p_j, p_k)$] with the full model. Results in Fig. 2.7 demonstrate that, without the interdependencies, the no-IPC model cannot accurately infer phrases from noisy motion attributes by the feature function $\phi_3(p_j, a_{j(i)})$. For example, in Fig. 2.7a, b, the phrases of occluded legs cannot be detected. However, the phrases of legs play key roles in recognizing "boxing" and "pat" [see Fig. 2.8 (right)]. Without the key phrases, the interactions cannot be recognized. By comparison, the full model can use the feature function $\phi_4(p_j, p_k)$ to learn the interdependencies of a pair of interactive phrases from training data. Once some phrases cannot be inferred from the corresponding attributes, the interdependencies will play a strong prior on the phrases and thus facilitate the recognition task.

Table 2.1 Accuracies of our method and action co-occurrence based method [23]

Methods	Lan et al. [23]	**Our method**
Accuracy	82.21 %	**85.16 %**

Table 2.2 Comparison results of accuracy (%) on the BIT-interaction dataset

Methods	Overall	Bow	Boxing	Handshake	High-five	Hug	Kick	Pat	Push
Bag-of-words	70.31	81.25	75	50	75	81.25	68.75	62.5	68.75
No-phrase method	73.43	81.25	68.75	68.75	81.25	68.75	81.25	62.5	75
No-IPC method	80.47	81.25	68.75	81.25	87.5	81.25	81.25	75	87.5
No-AC method	81.25	81.25	62.5	81.25	87.5	93.75	81.25	81.25	81.25
Our method	**85.16**	81.25	81.25	81.25	93.75	93.75	81.25	81.25	87.5

Interactive phrases have different importance to an interaction class. We illustrate the learned importance of interactive phrases to eight interaction classes in Fig. 2.8 (left). This figure demonstrates that our model learns some key interactive phrases to an interaction class (e.g., "interaction between embracing arms" in "hug" interaction). As long as these key interactive phrases are correctly detected, an interaction can be easily recognized. The learned top three key interactive phrases in all interaction classes are displayed in Fig. 2.8 (right).

We also compare our description-based method with action co-occurrence based method [23] for human interaction recognition. To conduct a fair comparison, we use the same bag-of-words motion representation for the two methods. Results in Table 2.1 indicate that our method outperforms the action co-occurrence based method. The underlying reason is that our method decomposes action co-occurrence relationships into a set of phrase co-occurrence relationships. The compositions of binary phrase variables allow us to represent interaction classes with large variations and thus make our method more expressive than [23].

Contributions of Components In this experiment, we evaluate the contributions of components in the proposed method, including the interdependencies in the attribute model and the interaction model, respectively, and the interactive phrases. We remove these components from our method, respectively, and obtain three different methods: the method without connections between attributes (no-AC method), the method without connections between interactive phrases (no-IPC method), and the interaction model without phrases (no-phrase method). Our method is compared with these three methods as well as the baseline bag-of-words representation with a linear SVM classifier.

Table 2.2 indicates that our method outperforms all the baseline methods. Compared with the baseline bag-of-words method, the performance gain achieved by our method is significant due to the use of high-level knowledge of human interaction. Our method significantly outperforms the no-phrase method, which demonstrates the effectiveness of the proposed interactive phrases. Our method

Fig. 2.9 Results of our method on UT-interaction dataset. In (**b**), correctly recognized examples are in the *first three columns* and misclassifications are in the *last column*. (**a**) Confusion matrix. (**b**) Classification examples of our method

uses interactive phrases to better represent complex human interactions and thus achieves superior results. As expected, the results of the proposed method are higher than the no-IPC method and the no-AC method, which emphasize the importance of the interdependencies between interactive phrases and attributes, respectively. With the interdependencies, the proposed method can capture the co-occurrences of interactive phrases and thus reduces the number of incorrect interactive phrases. The interdependencies between individual attributes enable to capture the important relationships between individual attributes and reduce inaccurate attribute labels caused by noisy features and subjective attribute labeling. With the interdependencies in both attribute pairs and interactive phrase pairs, our method can recognize some challenging interaction videos and thus achieves higher results.

2.5 Conclusion

We have proposed interactive phrases, semantic descriptions of motion relationships between people, for human interaction recognition. Interactive phrases incorporate rich human knowledge and thus provide an effective way to represent complex interactions. We have presented a novel method to encode interactive phrases, which is composed of the attribute model and the interaction model. Extensive experiments have been conducted and showed the effectiveness of the proposed method.

The attributes and phrases rely on expert knowledge and are dataset specific. Scaling up attributes and phrases to general datasets remains an open problem. Possible solutions are: (1) cross-dataset techniques and (2) data-driven attributes. Structure learning techniques can also be adopted to adaptively determine the optimal connectivity pattern between attributes and phrases in new datasets. We plan to explore these in future work.

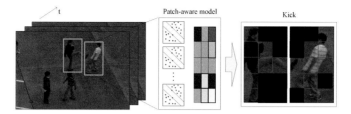

Fig. 2.10 Example of the inference results of our patch-aware model. Our model recognizes human interaction and discriminatively learns the supporting regions for each interacting people

3 Approach II: Human Interaction Recognition by Learning Discriminative Spatiotemporal Patches

In this chapter, we propose a novel patch-aware model [17] for solving the afore-mentioned problems in close human interaction recognition from videos (Fig. 2.10). Our model learns discriminative supporting regions for each interacting person, which accurately separate the target person from background. The learned supporting regions also indicate the feature-to-person assignments, which consequently help better represent individual actions. In addition, each interaction class associates with a variety of supporting region configurations, thereby providing rich and robust representations for different occlusion cases.

We propose a rich representation for close interaction recognition. Specifically, we introduce a set of binary latent variables for 3D patches indicating which subject the patch is associated with (background, person 1, or person 2), and encourage consistency of the latent variables across all the training data. The appearance and structural information of patches is jointly captured in our model, which captures the motion and pose variations of interacting people. To address the challenge of an exponentially large label space, we use a structured output framework, employing a latent SVM [10]. During training, the model learns which patterns belong to the foreground and background, allowing for better labeling of body parts and identification of individual people. Results show that the learned supporting patches significantly facilitate the recognition task.

3.1 Interaction Representation

Our approach takes advantage of 3D spatiotemporal local features to jointly recognize interaction and segment people in the interaction. Given a video, a visual tracker is applied to extract interacting people from each other, and also differentiate them from the background at a patch-level. In each bounding box, spatiotemporal interest points [8] and tracklet [36] are computed within each 3D patch, and described using the bag-of-words model [8, 25, 28] (Fig. 2.11). Spatiotemporal

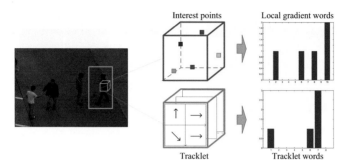

Fig. 2.11 Illustration of feature representation. We extract both interest points and tracklet from 3D patches

patches are obtained by decomposing a video of size $R \times C \times T$ into a set of non-overlapping spatiotemporal 3D patches, each of which is of size $r \times c \times t$. Similar to action representation based on histograms of video words [8, 28, 37], we describe each patch by the histogram of video words within the patch.

Noted that the detected interest points and tracklet are mainly associated with salient regions in human body; few of them are associated with background. This results in an inexpressive representation for background. Our aim in this chapter is to extract each interacting people from the interactions and thus the background must be described. In this chapter, we augment *virtual video words* (VVWs) to describe background.

The idea of VVWs is to build a discriminative feature for background so that background and foreground can be well differentiated. Consider the features of patches as data points in a high-dimensional space. Then patch features associated with foreground are distributed subjecting to an unknown probability. We would like to define some virtual data points for background and make them as far as possible from those foreground data points in order to make these two-class data points well separated. Since we use linear kernel in the model, the best choice for virtual data points is the one that can be linearly separated from foreground data points. In our work, we use origin point for virtual data points, i.e. all the bins in the histogram of a 3D patch which have no video words in it are set to 0.

3.2 Patch-Aware Model

Given the representation of an interaction video, our goal is to determine the interaction class (e.g., "push") as well as infer supporting regions for each interacting person. These 3D regions in this work can be associated with background or one of the interacting people.

Suppose we are given N training samples $\{\mathbf{x}^{(i)}, y^{(i)}\}_{i=1}^{N}$, where $\mathbf{x} \in \mathbb{R}^D$ denotes the video feature and $y \in \mathcal{Y}$ is the interaction class. Our purpose is to learn a discriminative function $f_w : \mathbf{x} \rightarrow y$, which infers the interaction class for an unknown interaction video. To model the supporting regions for each interacting person, we introduce a set of auxiliary binary latent variables $\{h_j\}_{j=1}^M \in \mathcal{H}$ ($h_j \in \{0, 1\}$), each of which associates with one patch. $h_j = 0$ denotes that the jth patch is associated with the background and $h_j = 1$ means it is with foreground. Note that intra-class variability leads to different patch configurations in certain interaction classes. For instance, in "handshake," some people would like to pat the other people while shaking hands with the people but some do not like that. We solve this problem by treating regions as latent variables and inferring the most probable states of latent variables in training. An undirected graph $\mathcal{G} = (\mathcal{V}, \mathcal{E})$ is employed to encode the configurations of these patches. A vertex $h_j \in \mathcal{V}$ ($j = 1, \ldots, M$) corresponds to the jth patch and an edge $(h_j, h_k) \in \mathcal{E}$ corresponds to the dependency between the two patches.

We define the discriminative function as

$$f(\mathbf{x}; \mathbf{w}) = \arg\max_y \left[\max_{\mathbf{h}} F(\mathbf{x}, \mathbf{h}, y; \mathbf{w}) \right], \quad (2.11)$$

where $\mathbf{h}$ is the vector of all latent variables. The scoring function $F(\mathbf{x}, \mathbf{h}, y; \mathbf{w})$ is used to measure the compatibility between the video data $\mathbf{x}$, the interaction class y and the latent patch labels $\mathbf{h}$.

We model the scoring function $F(\cdot)$ as a linear function $F(\mathbf{x}, \mathbf{h}, y; \mathbf{w}) = \langle \mathbf{w}, \Phi(\mathbf{x}, \mathbf{h}, y) \rangle$ with $\mathbf{w}$ being model parameter and $\Phi(\mathbf{x}, \mathbf{h}, y)$ being a feature vector. Specifically, the scoring function $F(\cdot)$ is defined as the summation of four components:

$$F(\mathbf{x}, \mathbf{h}, y; \mathbf{w}) = \sum_{j \in \mathcal{V}} \alpha^T \psi(\mathbf{x}_j, h_j, y) + \sum_{j \in \mathcal{V}} \beta^T \theta(\mathbf{x}_j, h_j)$$
$$+ \sum_{j \in \mathcal{V}} \gamma_j^T \eta(h_j, y) + \lambda^T \pi(\mathbf{x}, y), \quad (2.12)$$

where $\mathbf{w} = \{\alpha, \beta, \gamma, \lambda\}$ is the model parameter, $\mathbf{x}_j$ is the feature extracted from the jth patch.

Class-Specific Patch Model $\alpha^T \psi(\mathbf{x}_j, h_j, y)$ models the agreement between the observed patch feature $\mathbf{x}_j$, the patch label h_j and the interaction class y. The definition of the feature vector $\psi(\mathbf{x}_j, h_j, y)$ is given by

$$\psi(\mathbf{x}_j, h_j, y) = \mathbf{1}(y = a) \cdot \mathbf{1}(h_j = b) \cdot f(\mathbf{x}_j), \quad (2.13)$$

where $f(\mathbf{x}_j)$ denotes the local feature of the jth patch and $\mathbf{1}(\cdot)$ is an indicator function. In our work, $f(\mathbf{x}_j)$ encodes both appearance information and structural information of the jth patch: $f(\mathbf{x}_j) = [f_a(\mathbf{x}_j), f_s(\mathbf{x}_j)]$. The appearance information $f_a(\mathbf{x}_j)$ is the distribution of words in the patch, and the structural information $f_s(\mathbf{x}_j)$ is the location of the patch. To compute the structural feature $f_s(\mathbf{x}_j)$, we discretize the bounding box into M patches and the spatial location feature of a patch $\mathbf{x}_j$ can be represented as a vector of all zeros with a single 1 for the bin occupied by $\mathbf{x}_j$. We apply a template α of size $(D + M) \times H \times Y$ on the feature function $\psi(\mathbf{x}_j, h_j, y)$ to weigh the different importance of elements in the feature function, where Y is the number of interaction classes, and H is the number of patch labels. Each entry in α_{yhcm} can be interpreted as, for patch of state h, how much the proposed model prefers to see a discriminative word in the mth bin when the codeword is c and the interaction label is y. The class-specific patch model $\alpha^T \psi(\mathbf{x}_j, h_j, y)$ can be regarded as a linear classifier and scores the feature vector $\psi(\mathbf{x}_j, h_j, y)$.

The model encodes class-specific discriminative patch information which is of great importance in recognition. Note that the patch label h is unobserved during training and the feature function defined above models the implicit relationship between an interaction class and supporting regions. During training, the model automatically "aware" the supporting regions for an interaction class by maximizing the score $F(\mathbf{x}, \mathbf{h}, y; \mathbf{w})$.

Global Patch Model $\beta^T \theta(\mathbf{x}_j, h_j)$ measures the compatibility between the observed patch feature $\mathbf{x}_j$ and the patch label h_j. We define the feature function $\theta(\mathbf{x}_j, h_j)$ as

$$\theta(\mathbf{x}_j, h_j) = \mathbf{1}(h_j = b) \cdot f(\mathbf{x}_j), \tag{2.14}$$

where $f(\mathbf{x}_j)$ is the local feature of the jth patch used in the class-specific patch model. This model encodes shared patch information across interaction classes. It is a standard linear classifier trained to infer the label (0 or 1) of the jth patch given patch feature $\mathbf{x}_j$. The parameter β is a template, which can be considered as the parameter of a binary linear SVM trained with data $\{\mathbf{x}_j, h_j\}_{j=1}^M$.

Essentially, the global patch model encodes the shared patch information across interaction classes. For example, since we use a tracker to obtain a bounding box of an interacting person, this person tends to appear in the middle of the box and thus the patches in the middle of the box are likely to be labeled as foreground. This information is shared across all interaction classes and can be elegantly encoded by our global patch model.

Class-Specific Structure Model $\gamma_j^T \eta(h_j, y)$ encodes the structural information of patches in one interaction class. Intuitively, human poses are different in various interaction classes. Although this information are unobserved in training samples, we treat them as latent variables so that they can be automatically discovered during model training. The class-specific structure model is given by

$$\eta(h_j, y) = \mathbf{1}(h_i = b) \cdot \mathbf{1}(y = a). \tag{2.15}$$

Clearly, the label of a patch is related to its location. Therefore, we use a set of untied weights $\{\gamma\}_{j=1}^M$ for the jth patch, each of which is of size $H \times Y$, where M is the number of patches. The class-specific structure model expresses the prior that, without observing any feature, given an interaction class a, which state of the jth patch is likely to be.

The class-specific structure model expresses the idea that, without observing any low-level feature, given an interaction class a, which state of the jth patch is likely to be. The model shows its preference by scoring the feature vector $\eta(h_j, y)$ using a weight vector γ_j. Since the feature vector is a $0 - 1$ vector, if an entry in $\gamma_j(b, a)$ is positive, the model encourages labeling the jth patch as b when current interaction class is a.

Global Interaction Model $\lambda^T \pi(\mathbf{x}, y)$ is used to differentiate different interaction classes. We define this feature vector as

$$\pi(\mathbf{x_0}, y) = \mathbf{1}(y = a) \cdot \mathbf{x_0}, \qquad (2.16)$$

where $\mathbf{x}_0 \in \mathbb{R}^d$ is a feature vector extracted from the whole action video. Here we use the bag-of-words representation for the whole video. This potential function is essentially a standard linear model for interaction recognition if other components are not considered. If other potential functions in Eq. (2.12) are ignored, and only the global interaction potential function is considered, the parameter λ can be learned by a standard multi-class linear SVM.

Discussion The proposed patch-aware model is specifically designed for interaction recognition with close physical contact. Compared with existing interaction recognition methods [2, 3, 24, 33, 37–39, 45, 51], our model accounts for motion at a fine-grain patch level using the three components, the class-specific patch component, the global patch component, and the class-specific structure component. These three components model the appearance and structural information of local 3D patches and allow us to accurately separate interacting people at patch-level. To our best knowledge, our work is the first one that provides supporting patches for close interaction recognition, which can be used to separate interacting people.

3.3 Model Learning and Testing

Learning The latent SVM formulation is employed to train our model given the training examples $\mathcal{D} = \{\mathbf{x}^{(n)}, y^{(n)}\}_{n=1}^{N}$:

$$\min_{\mathbf{w}, \xi} \frac{1}{2} \|\mathbf{w}\|^2 + C \sum_n (\xi_n + \sigma_n) \qquad (2.17)$$

$$\text{s.t. } \max_{\mathbf{h}} \mathbf{w}^T \Phi(\mathbf{x}^{(n)}, \mathbf{h}_{y^{(n)}}, y^{(n)}) - \max_{\mathbf{h}} \mathbf{w}^T \Phi(\mathbf{x}^{(n)}, \mathbf{h}, y) \qquad (2.18)$$

$$\geq \Delta(y, y^{(n)}) - \xi_n, \forall n, \forall y,$$

$$\mu(\mathbf{h}_{y^{(n)}}, y^{(n)}, \mathbf{h}_y, y) \leq \sigma_n, \forall n, \forall y, \qquad (2.19)$$

where $\mathbf{w}$ denotes model parameter, ξ and σ are slack variables that allow for soft margin, and C is the soft-margin parameter. $\Delta(y, y^{(n)})$ represents the 0-1 loss function. $\mu(\mathbf{h}_{y^{(n)}}, y^{(n)}, \mathbf{h}_y, y)$ in Constraint (2.19) enforces the similarity over latent regions for training videos. Our assumption is that, for videos in the same category, they are likely to have the same latent variable values. We define $\mu(\mathbf{h}_{y^{(n)}}, y^{(n)}, \mathbf{h}_y, y)$ as

$$\mu(\mathbf{h}_{y^{(n)}}, y^{(n)}, \mathbf{h}_y, y) = \frac{1}{M} d(\mathbf{h}_{y^{(n)}}, \mathbf{h}_y) \cdot \mathbf{1}(y = y^{(n)}), \qquad (2.20)$$

where $d(\cdot, \cdot)$ computes the Hamming distance between the two vectors. The optimization problem (2.17)–(2.19) can be solved using the latent SVM framework [10].

Computing Subgradient The above optimization problem can be efficiently solved by the non-convex cutting plane algorithm [7]. The key idea of this algorithm is that it iteratively approximates the objective function by increasingly adding new cutting planes to the quadratic approximation. The two major steps of the algorithm are to compute the empirical loss $R(\mathbf{w}) = \sum_n (\xi_n + \sigma_n)$ and the subgradient $\frac{\partial R}{\partial \mathbf{w}}$.

The computation of a subgradient is relatively straightforward, assuming the inference over $\mathbf{h}$ can be done. Denote the empirical loss $R(\mathbf{w})$ as $R(\mathbf{w}) = \sum_n R^n(\mathbf{w})$, then the subgradient can be computed by

$$\frac{\partial R}{\partial \mathbf{w}} = \Phi(\mathbf{x}^{(n)}, \mathbf{h}^*, y^*) - \Phi(\mathbf{x}^{(n)}, \mathbf{h}', y^{(n)}), \qquad (2.21)$$

where $(\mathbf{h}^*, y^*)$ and $\mathbf{h}'$ are computed by

$$(\mathbf{h}^*, y^*) = \arg\max_{y, \mathbf{h}} \mathbf{w}^{\mathsf{T}} \Phi(\mathbf{x}^{(n)}, \mathbf{h}, y) + \Delta(y^{(n)}, y), \qquad (2.22)$$

$$\mathbf{h}' = \arg\max_{\mathbf{h}} \mathbf{w}^{\mathsf{T}} \Phi(\mathbf{x}^{(n)}, \mathbf{h}, y^{(n)}) - \mu(\mathbf{h}_{y^{(n)}}, y^{(n)}, \mathbf{h}, y). \qquad (2.23)$$

Testing Given an unknown interaction video, we assume that the interaction region in the video is known. Our aim is to infer the optimal interaction label y^* and the optimal configurations of 3D patches $\mathbf{h}^*$:

$$\max_y \max_{\mathbf{h}} \mathbf{w}^{\mathsf{T}} \Phi(\mathbf{x}, \mathbf{h}, y). \qquad (2.24)$$

To solve the above optimization problem, we enumerate all possible interaction classes $y \in \{\mathcal{Y}\}$ and solve the following optimization problem:

$$\mathbf{h}_y^* = \arg\max_{\mathbf{h}} \mathbf{w}^{\mathsf{T}} \Phi(\mathbf{x}, \mathbf{h}, y), \forall y \in \mathcal{Y}. \qquad (2.25)$$

Here, the latent variables $\mathbf{h}$ are connected by a lattice. In this work, we adopt loopy belief propagation to solve the above optimization problem.

Given the latent variable vector $\mathbf{h}_y^*$, we then compute the score $f_\mathbf{w}(\mathbf{x}, \mathbf{h}_y^*, y) = \mathbf{w}^\mathrm{T}\Phi(\mathbf{x}, \mathbf{h}_y^*, y)$ for all interaction classes $y \in \mathcal{Y}$ and pick up the optimal interaction class y^* which maximizes the score $F(\mathbf{x}, \mathbf{h}_y^*, y; \mathbf{w})$.

3.4 Experiments

3.4.1 Datasets

We test our method on the UT-Interaction dataset [40]. UT dataset consists of six classes of human interactions: handshake, hug, kick, point, punch, and push. The UT dataset was recorded for the human activity recognition contest (SDHA 2010) [40], and it has been used by several state-of-the-art action recognition methods [37, 39, 51].

3.4.2 Experiment Settings

We extract 300 interest points [8] from a video on the UT dataset. Gradient descriptors are utilized to characterize the motion around interest points. Principal component analysis algorithm is applied to reduce the dimensionality of descriptors to 100 and build a visual word vocabulary of size 1000. We use a visual tracker to obtain a bounding box for each interacting people. Then a 3D volume computed by stacking bounding boxes along temporal axis is split into non-overlapping spatiotemporal cuboids of size $15 \times 15 \times 15$. We use the histogram of the video words in a 3D patch as the patch feature. We adopt the leave-one-out training strategy on the UT dataset.

3.4.3 Results on UT-Interaction Dataset

On UT dataset, we first evaluate the recognition accuracy of our method and report supporting region results. Then we compare with state-of-the-art methods [19, 24, 28, 37, 51].

Recognition Accuracy We test our method on UT dataset and show the confusion matrix in Fig. 2.12. Our method achieves 88.33 % recognition accuracy. Confusions are mainly due to visually similar movements in two classes (e.g., "push" and "punch") and the influence of moving objects in the background. Classification examples are illustrated in Fig. 2.12.

Equation (2.15) defines a class-specific structure model for all classes. It would be interesting to investigate the performance of a shared pose prior. We replace the class-specific structure prior in Eq. (2.15) with a shared one which is defined as $\eta(h_j, y) = 1(h_i = b)$. Results are shown in Table 2.3. The accuracy difference

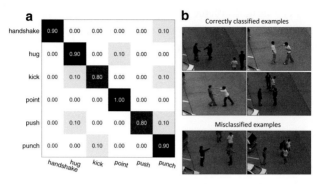

Fig. 2.12 Confusion matrix and classification examples of our method on UT dataset. (**a**) Confusion matrix. (**b**) Predicted examples

Table 2.3 Accuracies of different pose prior on UT dataset

Pose prior	Shared	Class-specific
Accuracy	83.33 %	88.33 %

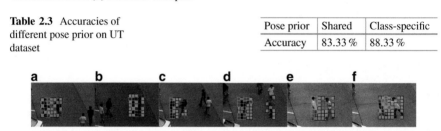

Fig. 2.13 The learned supporting regions on the UT dataset. (**a**) Handshake. (**b**) Hug. (**c**) Kick. (**d**) Point. (**e**) Punch. (**f**) Push

between the two priors is 5 %. This is mainly due to that motion variations in individual actions are significant. The model with class-specific prior is able to learn pose under different classes, and benefits the recognition task.

Supporting Regions The learned supporting regions on the UT dataset are shown in Fig. 2.13. Our model can accurately discover supporting regions of interacting people. This is achieved by finding the most discriminative regions (e.g. hand and leg) that support an interaction class. Note that some videos in the UT dataset have background motion, e.g.,"point," which introduces noise in the video. However, our model uses the structure prior component in Eq. (2.15) and the consistency Constraint (2.19) to enforce a strong structure prior information on the patches, and thus can determine which patches are unlikely to be associated with foreground. This leads to accurate patch labeling results. Some of the patch labels are incorrect mainly due to intra-class variations. People in an interaction class may behave differently according to their personal habits. This increases the difficulty of learning class-specific pose prior.

Comparison Results We evaluate the value of components in the proposed model, including the global interaction model, the structure prior model, and the patch models. We remove these from our patch-aware model respectively, and obtain

Table 2.4 Recognition accuracy (%) of methods on the UT dataset

Methods	Function	Handshake	Hug	Kick	Point	Punch	Push	Overall
Bag-of-words	Only rec.	70	70	80	90	70	70	75
No-GI method	Rec. and seg.	20	30	40	30	10	20	25
No-SP method	Rec. and seg.	70	80	70	70	80	80	75
No-CGP method	Rec. and seg.	80	90	70	90	80	80	81.67
Liu et al. [28]	Only rec.	60	70	100	80	60	70	73.33
Lan et al. [24]	Only rec.	70	80	80	80	90	70	78.33
Yu et al. [51]	Only rec.	100	65	75	100	85	75	83.33
Ryoo and Aggarwal [37]	Only rec.	80	90	90	80	90	80	85
Our method	Rec. and seg.	90	90	80	100	80	90	**88.33**

three different methods: the no-GI method that removes global interaction potential $\lambda^{\mathrm{T}}\pi(\mathbf{x}, y)$, the no-SP method that removes the structure prior potential $\gamma_j^{\mathrm{T}}\eta(h_j, y)$, and the no-CGP method which removes both class-specific and global patch model $\alpha^{\mathrm{T}}\psi(\mathbf{x}_j, h_j, y)$ and $\beta^{\mathrm{T}}\theta(\mathbf{x}_j, h_j)$ from the full model.

We compare our full model with previous methods [24, 28, 37, 51], the no-GI method, no-SP method, and no-CGP method, and adopt a bag-of-words representation with a linear SVM classifier as the baseline. Results in Table 2.4 show that our method outperforms all the comparison methods. It should be noted that our method learns supporting regions, which can be used to separate people while the methods in [24, 28, 37, 51] cannot achieve this goal.

Results in Table 2.4 show that our method outperforms [24, 28, 37, 51]. The baseline bag-of-words method simply uses low-level features for recognition. By comparison, our method treats cuboid variables as mid-level features and utilize them to describe local motion information. With rich representation of interaction, our method achieves superior performance. Our method outperforms the method proposed in [37]. Their method uses structural information between interest points to aid recognition. In this work, we adopt a different scheme to encode structure information of interest points. The information is encoded by the location of spatiotemporal cuboids which contains the interest points. Besides, the learned supporting regions in our model can also be used to separate people in interactions while their method cannot. Lan et al. [24] utilized action context to recognize interactions. We argue that action context may not able to capture complex action co-occurrence since individual motion could be totally different in an interaction class. Thus modeling the action context may not capture significant motion variations in individual actions. We infer an interaction based on the mid-level patch features. The mid-level features we build can provide detailed regional motion information of interactions and thus improve recognition results. Compared with [51], our method learns supporting regions to separate people while [51] treats interacting people as a group and do not consider separation.

3.5 Summary

We have proposed a novel model for jointly recognizing human interaction and segmenting people in the interaction. Our model is built upon the latent structural support vector machine in which the patches are treated as latent variables. The consistency of latent variables is encouraged across all the training data. The learned patch labels indicate the supporting regions for interacting people, and thus solve the problems of feature assignment and occlusion. Experiments show that our method achieves promising recognition results and can segment people at patch level during an interaction, even in a close interaction.

References

1. Aggarwal, J.K., Ryoo, M.S.: Human activity analysis: a review. ACM Comput. Surv. **43**(3), pp. 16:1–16:43. (2011)
2. Choi, W., Savarese, S.: A unified framework for multi-target tracking and collective activity recognition. In: European Conference on Computer Vision, pp. 215–230. Springer, Berlin (2012)
3. Choi, W., Shahid, K., Savarese, S.: Learning context for collective activity recognition. In: Conference on Computer Vision and Pattern Recognition (2011)
4. Chow, C.K., Liu, C.N.: Approximating discrete probability distributions with dependence tree. IEEE Trans. Inf. Theory **14**(3), 462–467 (1968)
5. Desai, C., Ramanan, D., Fowlkes, C.: Discriminative models for multi-class object layout. In: International Conference on Computer Vision (2009)
6. Desai, C., Ramanan, D., Fowlkes, C.: Discriminative models for static human-object interactions. In: Conference on Computer Vision and Pattern Recognition Workshop on Structured Models in Computer Vision (2010)
7. Do, T.-M.-T., Artieres, T.: Large margin training for hidden Markov models with partially observed states. In: International Conference on Machine Learning (2009)
8. Dollar, P., Rabaud, V., Cottrell, G., Belongie, S.: Behavior recognition via sparse spatio-temporal features. In: Visual Surveillance and Performance Evaluation of Tracking and Surveillance (2005)
9. Farhadi, A., Endres, I., Hoiem, D., Forsyth, D.: Describing objects by their attributes. In: Conference on Computer Vision and Pattern Recognition, pp. 1778–1785. IEEE (2009)
10. Felzenszwalb, P., McAllester, D., Ramanan, D.: A discriminatively trained, multiscale, deformable part model. In: Conference on Computer Vision and Pattern Recognition (2008)
11. Ferrari, V., Zisserman, A.: Learning visual attributes. In: Conference on Neural Information Processing Systems (2007)
12. Filipovych, R., Ribeiro, E.: Recognizing primitive interactions by exploring actor-object states. In: Conference on Computer Vision and Pattern Recognition, pp. 1–7. IEEE, New York (2008)
13. Gong, S., Xiang, T.: Recognition of group activities using dynamic probabilistic networks. In: International Conference on Computer Vision, vol. 2, pp. 742–749 (2003)
14. Gorelick, L., Blank, M., Shechtman, E., Irani, M., Basri, R.: Actions as space-time shapes. IEEE Trans. Pattern Anal. Mach. Intell. **29**(12), 2247–2253 (2007)
15. Gupta, A., Davis, L.S.: Beyond nouns: exploiting prepositions and comparative adjectives for learning visual classifiers. In: European Conference on Computer Vision (2008)
16. Gupta, A., Kembhavi, A., Davis, L.S.: Observing human-object interactions: using spatial and functional compatibility for recognition. IEEE Trans. Pattern Anal. Mach. Intell. **31**(10), 1775–1789 (2009)

17. Kong, Y., Fu, Y.: Modeling supporting regions for close human interaction recognition. In: European Conference on Computer Vision Workshop (2014)
18. Kong, Y., Jia, Y., Fu, Y.: Learning human interaction by interactive phrases. In: European Conference on Computer Vision (2012)
19. Kong, Y., Jia, Y., Fu, Y.: Interactive phrases: semantic descriptions for human interaction recognition. IEEE Trans. Pattern Anal. Mach. Intell. **36**, 1775–1788 (2014)
20. Kovashka, A., Grauman, K.: Learning a hierarchy of discriminative space-time neighborhood features for human action recognition. In: Conference on Computer Vision and Pattern Recognition, pp. 2046–2053. IEEE, New York (2010)
21. Lafferty, J., McCallum, A., Pereira, F.: Conditional random fields: probabilistic models for segmenting and labeling sequence data. In: International Conference on Machine Learning (2001)
22. Lampert, C.H., Nickisch, H., Harmeling, S.: Learning to detect unseen object classes by between-class attribute transfer. In: Conference on Computer Vision and Pattern Recognition (2009)
23. Lan, T., Wang, Y., Yang, W., Mori, G.: Beyond actions: discriminative models for contextual group activities. In: Conference on Neural Information Processing Systems (2010)
24. Lan, T., Wang, Y., Yang, W., Robinovitch, S.N., Mori, G.: Discriminative latent models for recognizing contextual group activities. IEEE Trans. Pattern Anal. Mach. Intell. **34**(8), 1549–1562 (2012)
25. Laptev, I., Marszałek, M., Schmid, C., Rozenfeld, B.: Learning realistic human actions from movies. In: Conference on Computer Vision and Pattern Recognition (2008)
26. Li, R., Chellappa, R., Zhou, S.K.: Learning multi-modal densities on discriminative temporal interaction manifold for group activity recognition. In: Conference on Computer Vision and Pattern Recognition, pp. 2450–2457 (2009)
27. Liu, J., Luo, J., Shah, M.: Recognizing realistic actions from videos "in the wild". In: Proceedings of IEEE Conference on Computer Vision and Pattern Recognition (2009)
28. Liu, J., Kuipers, B., Savarese, S.: Recognizing human actions by attributes. In: Conference on Computer Vision and Pattern Recognition (2011)
29. Marszałek, M., Laptev, I., Schmid, C.: Actions in context. In: Proceedings of IEEE Conference on Computer Vision and Pattern Recognition (2009)
30. Mehran, R., Oyama, A., Shah, M.: Abnormal crowd behavior detection using social force model. In: Conference on Computer Vision and Pattern Recognition (2009)
31. Ni, B., Yan, S., Kassim, A.A.: Recognizing human group activities with localized causalities. In: Conference on Computer Vision and Pattern Recognition, pp. 1470–1477 (2009)
32. Niebles, J.C., Chen, C.-W., Fei-Fei, L.: Modeling temporal structure of decomposable motion segments for activity classification. In: European Conference on Computer Vision, vol. 6312 (2010)
33. Odashima, S., Shimosaka, M., Kaneko, T., Fuikui, R., Sato, T.: Collective activity localization with contextual spatial pyramid. In: European Conference on Computer Vision (2012)
34. Oliver, N.M., Rosario, B., Pentland, A.P.: A Bayesian computer vision system for modeling human interactions. IEEE Trans. Pattern Anal. Mach. Intell. **22**(8), 831–843 (2000)
35. Patron-Perez, A., Marszalek, M., Reid, I., Zissermann, A.: Structured learning of human interaction in tv shows. IEEE Trans. Pattern Anal. Mach. Intell. **34**(12), 2441–2453 (2012)
36. Raptis, M., Soatto, S.: Tracklet descriptors for action modeling and video analysis. In: European Conference on Computer Vision (2010)
37. Ryoo, M.S.: Human activity prediction: early recognition of ongoing activities from streaming videos. In: International Conference on Computer Vision (2011)
38. Ryoo, M.S., Aggarwal, J.K.: Recognition of composite human activities through context-free grammar based representation. In: Conference on Computer Vision and Pattern Recognition, vol. 2, pp. 1709–1718 (2006)
39. Ryoo, M.S., Aggarwal, J.K.: Spatio-temporal relationship match: video structure comparison for recognition of complex human activities. In: International Conference on Computer Vision, pp. 1593–1600 (2009)

40. Ryoo, M.S., Aggarwal, J.K.: UT-interaction dataset, ICPR contest on semantic description of human activities (SDHA) (2010). http://cvrc.ece.utexas.edu/SDHA2010/Human_Interaction. html
41. Ryoo, M., Aggarwal, J.: Stochastic representation and recognition of high-level group activities. Int. J. Comput. Vis. **93**, 183–200 (2011)
42. Sadeghi, M.A., Farhadi, A.: Recognition using visual phrases. In: Conference on Computer Vision and Pattern Recognition (2011)
43. Shechtman, E., Irani, M.: Space-time behavior based correlation. In: Conference on Computer Vision and Pattern Recognition, vol. 1, pp. 405–412. IEEE (2005)
44. Taskar, B., Guestrin, C., Koller, D.: Max-margin Markov networks. In: Conference on Neural Information Processing Systems (2003)
45. Vahdat, A., Gao, B., Ranjbar, M., Mori, G.: A discriminative key pose sequence model for recognizing human interactions. In: International Conference on Computer Vision Workshops, pp. 1729–1736 (2011)
46. Wang, Y., Mori, G.: Max-margin hidden conditional random fields for human action recognition. In: Conference on Computer Vision and Pattern Recognition, pp. 872–879 (2009)
47. Wang, Y., Mori, G.: A discriminative latent model of object classes and attributes. In: European Conference on Computer Vision (2010)
48. Wang, Y., Mori, G.: Hidden part models for human action recognition: probabilistic vs. max-margin. IEEE Trans. Pattern Anal. Mach. Intell. vol 33, pp. 1310–1323. (2010)
49. Yao, B., Fei-Fei, L.: Modeling mutual context of object and human pose in human-object interaction activities. In: Conference on Computer Vision and Pattern Recognition, pp. 17–24 (2010)
50. Yao, B., Fei-Fei, L.: Recognizing human-object interactions in still images by modeling the mutual context of objects and human poses. IEEE Trans. Pattern Anal. Mach. Intell. **34**(9), 1691–1703 (2012)
51. Yu, T.-H., Kim, T.-K., Cipolla, R.: Real-time action recognition by spatiotemporal semantic and structural forests. In: British Machine Vision Conference (2010)

Chapter 3
Subspace Learning for Action Recognition

Chengcheng Jia and Yun Fu

1 Introduction

Recently human action recognition [4, 5, 8–10] has aroused widely attention for public surveillance system, elder service system, etc. However, the data captured by webcams are often high dimensional and usually contain noise and redundancy. So it is crucial to extract the meaning information by mitigating uncertainties for higher accuracy of recognition task. From this motivation, there are two topics we propose in this chapter: (1) select key frames from a video to remove noise and redundancy and (2) learn a subspace for dimensional reduction to reduce time complexity.

Sparse canonical correlation analysis (SCCA) selects some important variables by setting some coefficients zero, which is generally used for high-dimensional genetical selection [2, 12, 15, 20]. Inspired by this concept, we propose to employ the dual shrinking method for dimensional reduction tasks on the high-dimensional visual data, such as human action video, face with different factors and some textual data. We aim to select some key frames/variates from the original high-dimensional data while containing most useful information for recognition task. Nevertheless, there are three limitations with the SCCA framework:

C. Jia (✉)
Department of Electrical and Computer Engineering, Northeastern University,
360 Huntington Avenue, Boston, MA 02115, USA
e-mail: jia.ch@husky.neu.edu

Y. Fu
Department of Electrical and Computer Engineering and College of Computer and Information
Science (Affiliated), Northeastern University, 360 Huntington Avenue, Boston, MA 02115, USA
e-mail: yunfu@ece.neu.edu

© Springer International Publishing Switzerland 2016   49
Y. Fu (ed.), *Human Activity Recognition and Prediction*,
DOI 10.1007/978-3-319-27004-3_3

(a) The image vectorization brings the dimensional disaster problem, i.e., the number of variables is far larger than the number of samples, which is limited for variable selection because the lasso selects at most the smaller number of variables [23].
(b) The time complexity of vectorization is very high when calculating the variance matrix, because of the large number of variables.
(c) The high correlations between variables brings the collinearity problem, which leads to a large variance of the estimates and makes the solution devolutive.

We aim to find a sparse method that works well as SCCA, while overcoming the above problems, i.e., it should improve the performance of variable selection limited by (a) and (c) and save time against (b).

In order to alleviate the case of limitations (a) and (b), we employ a tensor representation for the high-dimensional data and learn a tensor subspace for dimensional reduction. A number of tensor subspace analysis methods [6, 11, 18, 19, 21, 22] have been presented for recognition-related tasks. In [18], they explore self-similarity of human skeletons tensors for action recognition. In [22], they extract HOG feature from an action tensor for recognition. In [11], the multi-linear PCA (MPCA) method extracts the principle components and preserves most of the energy of the tensor. The discriminant analysis with tensor representation (DATER) [21] and general tensor discriminant analysis (GTDA) [19] were proposed by maximizing a discriminant criterion in tensor subspace for dimensional reduction. In [6], a tensor subspace analysis-based CCA algorithm, i.e., TCCA, was introduced to efficiently learn the maximum correlations between pairwise tensors iteratively, which is applied to high-dimensional action data processing. Inspired by the tensorial representation, we propose a tensor model to preprocess the high-dimensional data, in order to save time and memory. Moreover, we employ the elastic net instead of lasso in the traditional SCCA to overcome the limitation (c). The elastic net includes lasso penalty and ridge penalty, which are used to control the degree of shrinkage and overcome the collinearity by decorrelation, respectively.

In this chapter, a dual shrinking method is proposed using elastic net in tensor subspace, which is called sparse tensor subspace learning (STSL), for dimensional reduction. STSL is used to deal with high-dimensional data, like human action video, face sequences, texture data for key frames/variates selection, and dimension reduction.

We transform the original data set to a big tensor, which is illustrated in Fig. 3.1. The STSL method performs alternating variable selection of the training and testing tensors. We employ the elastic net to control the degree of shrinkage and eliminate the collinearity caused by the high correlation. Then we boost STSL by performing SVD on the covariance and sparse coefficient vector. We have analyzed the properties of STSL, including the root mean squared error (RMSE), shrinkage degree, explained variance, and time complexity. The aim of STSL is to calculate the transformation matrices for dimensional reduction of tensor. The advantages of our methods are stated as follows:

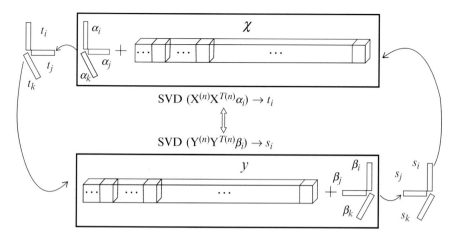

Fig. 3.1 Framework of dual shrinking in tensorial subspace. Firstly, the sparse loadings α_i, α_j, α_k of each mode has been estimated by elastic net regression; then the transformation vectors t_i, t_j, t_k has been calculated by performing SVD on the product of data covariance and sparse loadings; finally, the transformation vectors are used for the iterative sparse solution.

1. It can process high-dimensional data by tensorial representation.
2. The time cost decreases in the step of calculating the covariance.
3. The lasso is replaced by the elastic net which can perform shrinkage procedure and eliminate the collinearity problem through decorrelation.
4. The transformation matrices are not the ordinary linear combinations, but obtained by SVD on the product of covariance and the sparse coefficient vector.

The rest of this chapter is organized as follows: Section 2 introduces tensor and elastic net fundamentals. Section 3 presents the STSL algorithm and its properties. Experiments on common datasets are demonstrated in Section 4. Finally, the conclusion is given in Section 5.

2 Related Work and Motivation

2.1 Tensor Algebra

An *N-order* tensor can be represented as $\mathcal{A} \in \mathbb{R}^{I_1 \times I_2 \times \cdots \times I_n \times \cdots \times I_N}$, where I_n is the *mode-n* dimension of $\mathcal{A}$ ($1 \leqslant n \leqslant N$). An element of $\mathbf{A}$ is $\mathbf{A}_{:\ldots i_n \ldots:}$ ($1 \leq i_n \leq I_n$). The *mode-n* unfolding of $\mathcal{A}$ is denoted as matrix $\mathbf{A}^{(n)} \in \mathbb{R}^{I_n \times (I_1 \cdot I_2 \ldots I_{n-1} \cdot I_{n+1} \ldots I_N)}$, where $\times$ is used for measuring a matrix and $\cdot$ for scalar product. $\mathcal{A}$ is decomposed by $\mathbf{U}_n \in \mathbb{R}^{I_n \times J_n}$ ($1 \leqslant n \leqslant N$) as

$$\mathcal{B} = \mathcal{A} \times_1 \mathbf{U}_1^T \times_2 \mathbf{U}_2^T \ldots \times_N \mathbf{U}_N^T, \tag{3.1}$$

where $\mathcal{A} = \mathcal{B} \times_1 \mathbf{U}_1 \times_2 \mathbf{U}_2 \ldots \times_N \mathbf{U}_N$ and $\times_n$ indicates mode-*n* product. The transformed tensor $\mathcal{B} \in \mathbb{R}^{J_1 \times J_2 \times \cdots \times J_N}$ is called the core tensor of $\mathcal{A}$.

2.2   Loss Function

We intend to give the loss function as the converge condition. Given tensor $\mathcal{A}, \mathcal{B}$ as mentioned above, the loss function is defined as follows:

$$L(\mathcal{A}, \mathcal{B}) = |\,||\mathcal{A}||^2 - ||\mathcal{B}||^2\,|, \tag{3.2}$$

which means the absolute value of L_2-norm difference.

2.3   Regularized Objective Function

We employ lasso solution to select the important variates of data, and the trivial variates are set to zeros. Given two image sets $\mathbf{X}$ and $\mathbf{Y} \in \mathbb{R}^{m \times p}$, each of which has m samples with p variables. Given nonnegative lasso penalty λ_1, the lasso model [23] is obtained by

$$\hat{\beta} = \arg\min_{\beta} \|\mathbf{Y} - \mathbf{X}\beta\|^2 + \lambda_1 ||\beta||_1, \tag{3.3}$$

where $\hat{\beta} \in \mathbb{R}^p$ is a vector which consists of the lasso estimates, and $|\beta|_1 = \sum_{j=1}^{p} |\beta_j|$.

2.4   Optimization

The lasso method can pick up the significant variates for better representation. However, the lasso solution can bring in the singular problem, resulting in the large variance. We take a bias-variance trade-off solution, which is performed by combining ridge regression with lasso, to solve the problem. Given nonnegative ridge penalty λ_2, the elastic net model [23] is obtained by

$$\hat{\beta} = \arg\min_{\beta} \|\mathbf{Y} - \mathbf{X}\beta\|^2 + \lambda_1 ||\beta||_1 + \lambda_2 ||\beta||_1, \tag{3.4}$$

where $|\beta|^2 = \sum_{j=1}^{p} |\beta_j|^2$. The ridge solution has two advantages: (1) it can overcome the collinearity which causes the singular problem, and the step is called decorrelation; (2) it can avoid too sparse solution, by the rescaling factor $1/(1 + \lambda_2)$, which ensures the shrinkage step of the ridge is removed, maintaining the decorrelation effect of the ridge and shrinkage effect of the lasso. The solution of the elastic net is as follows:

$$\hat{\beta} = \arg\min_{\beta} \beta^T \left(\frac{\mathbf{X}^T\mathbf{X} + \lambda_2\mathbf{I}}{1 + \lambda_2} \right) \beta - 2y^T\mathbf{X}\beta + \lambda_1||\beta||_1, \qquad (3.5)$$

where $y \in \mathbb{R}^p$ is a sample. When $\lambda_2 = 0$, the elastic net turns to lasso. When $\lambda_2 \to \infty$, it results in the soft-thresholding [23, 24].

3 Sparse Tensor Subspace Learning

We represent each of the training and testing large data sets as a tensor and perform the minimizing variance procedure alternatively by fixing the parameters of one tensor while estimating those parameters of the other (Section 3.2). The selection of parameters is discussed in Section 3.3, following the time efficient is demonstrated in Section 4.5.3, respectively.

3.1 Motivation

We propose a sparse tensor-based method for high-dimensional data processing, to reduce dimension for recognition task. We aim to solve two problems: (1) select key frames by dual sparse learning to remove noise and redundancy and (2) learn a tensor subspace for dimensional reduction. The details are listed as follows:

1. The vector data is transformed to tensor representation, in order to reduce the dimension of data. Either the training or testing data set is represented to be a big tensor, as well as the vectorial method.
2. We use elastic net for sparse regression to select the significant variates of data. Lasso is used for sparsity, while ridge is used for overcoming the collinearity and alleviating double sparsity.
3. We employ a dual sparsity strategy, considering the correlation of the two data sets. However, it is not same to the traditional canonical correlation analysis. Firstly, we calculate the sparse coefficient loadings; then we alternatively update the transformation matrix by performing SVD on the product of data covariance and sparse coefficient [24], for the reason of obtaining the invariance eigenstructure of original data.

3.2 Learning Algorithm

In this part we detail the process of estimating the sparse loadings alternatively. Given two *N-order* tensors $\mathcal{X}$, $\mathcal{Y}$, we initialize the transformation matrices $\mathbf{S}_n$, $\mathbf{T}_n \in \mathbb{R}^{I_n \times J_n}$ $(1 \le n \le N)$ by *mode-n* eigen-decomposition of $\mathcal{X}$ and $\mathcal{Y}$, respectively. Firstly, $\mathcal{X}$, $\mathcal{Y}$ are dimensional reduced by

$$\mathcal{X} \leftarrow \mathcal{X} \times \mathbf{S}_1 \ldots \times_{n-1} \mathbf{S}_{n-1} \times_{n+1} \mathbf{S}_{n+1} \ldots \times \mathbf{S}_N, \tag{3.6}$$

$$\mathcal{Y} \leftarrow \mathcal{Y} \times \mathbf{T}_1 \ldots \times_{n-1} \mathbf{T}_{n-1} \times_{n+1} \mathbf{T}_{n+1} \ldots \times \mathbf{T}_N. \tag{3.7}$$

Secondly, $\mathcal{X}$ and $\mathcal{Y}$ are *mode-n* unfolded to $\mathbf{X}^{(n)}$ and $\mathbf{Y}^{(n)} \in \mathbb{R}^{I_n \times L_n}$, where $L_n = (J_1 \cdot J_2 \ldots J_{n-1} \cdot J_{n+1} \ldots J_N)$, both of which are centered. Initialized $\mathbf{A}^{(n)} = \{\alpha_j^{(n)}\}$, $\mathbf{B}^{(n)} = \{\beta_j^{(n)}\}$, $\mathbf{S}_n = \{s_j^{(n)}\}$, $\mathbf{T}_n = \{t_j^{(n)}\}$ $(1 \leq j \leq J_n)$, and the subscripts of $\mathbf{S}_n$ and $\mathbf{T}_n$ are dropped for simplicity. We use variance of sample, e.g., $\mathbf{X}^{(n)}\mathbf{X}^{T(n)}$, to calculate the sparse coefficients. There are two reasons for the solution: (1) when $L_n \gg I_n$, it can reduce dimension to great extent using the covariance in the space $\mathbb{R}^{I_n \times I_n}$ for the sparse processing, resulting in lower time complexity; (2) the vectors of transformation matrices $s^{(n)}$, $t^{(n)}$ have the same dimension as the sparse coefficient vectors $\alpha^{(n)}$, $\beta^{(n)}$ in the iteration; therefore, it is convenient to calculate the matrices $\mathbf{S}_n$, $\mathbf{T}_n$ we need.

Our dual shrinking model is obtained by learning the following optimization problems alternately.

1. Given $s^{(n)}$, the optimized $\hat{\alpha}^{(n)}$ is estimated as

$$\hat{\alpha}^{(n)} = \arg\min_{\alpha^{(n)}} \|s^{(n)} - \mathbf{X}^{(n)}\mathbf{X}^{T(n)}\alpha^{(n)}\|^2 + \lambda_1 |\alpha^{(n)}|_1 + \lambda_2 ||\alpha^{(n)}||^2,$$

1subject to $\mathrm{Var}\left(\mathbf{X}^{(n)}\mathbf{X}^{T(n)}\alpha^{(n)}\right) = 1.$ \hfill (3.8)

We use the Lagrangian function to calculate the solution as

$$\hat{\alpha}^{(n)} = \arg\min_{\alpha^{(n)}} \alpha^{T(n)} \left(\frac{(\mathbf{X}^{(n)}\mathbf{X}^{T(n)})^2 + \lambda_2 \mathbf{I}}{1 + \lambda_2} \right) \alpha^{(n)} \tag{3.9}$$
$$-2y^{T(n)}(\mathbf{X}^{(n)}\mathbf{X}^{T(n)})\alpha^{(n)} + \lambda_1 ||\alpha^{(n)}||_1$$

2. We use Procrustes rotation [13] to update $\mathbf{A}^{(n)}$ and perform SVD as

$$\mathbf{X}^{(n)}\mathbf{X}^{T(n)}\mathbf{A}^{(n)} = \mathbf{U}^{(n)}\mathbf{D}^{(n)}\mathbf{V}^{T(n)}, \tag{3.10}$$

and then update $\mathbf{T}_n = \mathbf{U}^{(n)}\mathbf{V}^{T(n)}$.

3. Given $t^{(n)}$, the optimized $\hat{\beta}^{(n)}$ is estimated as

$$\hat{\beta}^{(n)} =$$
$$\arg\min_{\beta^{(n)}} \|t^{(n)} - \mathbf{Y}^{(n)}\mathbf{Y}^{T(n)}\beta^{(n)}\|^2 + \lambda_1 ||\beta^{(n)}||_1 + \lambda_2 ||\beta^{(n)}||^2, \tag{3.11}$$

subject to $\mathrm{Var}\left(\mathbf{Y}^{(n)}\mathbf{Y}^{T(n)}\beta^{(n)}\right)^T = 1.$

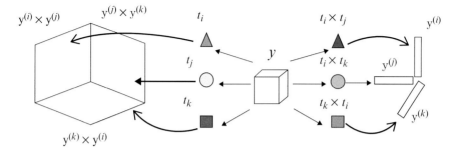

Fig. 3.2 Illustration of projection in tensor subspace. The different outer product of transformation vectors construct different projection subspace, by which the dimensional reduced tensor is obtained

The solution is as follow:

$$\hat{\beta}^{(n)} = \arg\min_{\beta^{(n)}} \beta^{T(n)} \left(\frac{(\mathbf{Y}^{(n)}\mathbf{Y}^{T(n)})^2 + \lambda_2 \mathbf{I}}{1 + \lambda_2} \right) \beta^{(n)} \tag{3.12}$$

$$-2x^{T(n)}(\mathbf{Y}^{(n)}\mathbf{Y}^{T(n)})\beta^{(n)} + \lambda_1 |\beta^{(n)}|_1,$$

4. update $\mathbf{B}^{(n)}$, then perform SVD as

$$\mathbf{Y}^{(n)}\mathbf{Y}^{T(n)}\mathbf{B}^{(n)} = \mathbf{U}'^{(n)}\mathbf{D}'^{(n)}\mathbf{V}'^{T(n)}, \tag{3.13}$$

then update $\mathbf{S}_n = \mathbf{U}'^{(n)}\mathbf{V}'^{T(n)}$.

After the iteration, we obtain the transformation matrices $\mathbf{S}_n$, $\mathbf{T}_n$, which are used for dimensional reduction. The projection process is shown in Fig. 3.2. Then the nearest neighborhood classification (NNC) is proposed. Algorithm 1 summarizes the proposed STSL in steps.

3.3 Selection of Parameters

In this part we will discuss how to select the parameters in the proposed method. We have defined (λ_1, λ_2) in Section 3.2, though it is not the only choice as parameters. For each fixed λ_2, we can also set the number of nonzero elements (Card) in the final sparse loadings as the lasso parameter, which is used to lighten the computational burden for early stopping [23]. In this chapter we use Card as the shrinkage parameter. The parameter Card is set initially and kept constant during the iterations.

Algorithm 1 Sparse Tensor Subspace Learning

INPUT: *N-order* training tensor set $\mathcal{X}$ and testing tensor set $\mathcal{Y}$. Given matrices $A^{(n)}, B^{(n)}$.
OUTPUT: Transformation matrices $\mathbf{S}_n, \mathbf{T}_n$ $(1 \leq n \leq N)$.
Algorithm:
Initialize $\mathbf{S}_n, \mathbf{T}_n \in \mathbb{R}^{I_n \times J_n}$ $(1 \leq n \leq N)$ by *mode-n* eigen-decomposition of $\mathcal{X}$ and $\mathcal{Y}$, respectively.
repeat
    for $n = 1$ to N **do**
        Dimensional reduction by Eqs. (3.6) and (3.7), *mode-n* unfolding to $\mathbf{X}^{(n)}, \mathbf{Y}^{(n)}$ and being centered.
        repeat
            for $i = 1$ to J_n **do**
                Given a fixed $s_i^{(n)}$, calculate $\alpha_i^{(n)}$ by Eq. (3.8), update matrix $\mathbf{T}_n$ by Eq. (3.10);
                Given a fixed $t_i^{(n)}$, calculate $\beta_i^{(n)}$ by Eq. (3.11), update matrix $\mathbf{S}_n$ by Eq. (3.13);
            end for
            Normalized updated $A'^{(n)}, B'^{(n)}$.
        until $|A'^{(n)} - A^{(n)}| < \epsilon$, $|B'^{(n)} - B^{(n)}| < \epsilon$.
    end for
until $L(\mathcal{Y}, \mathcal{Y}') = ||\mathcal{Y}|^2 - |\mathcal{Y}'|^2| < \epsilon'$, where $\mathcal{Y}'$ is the transformed tensor by the updated $\mathbf{T}_n$.

---

4 Experiment

4.1 Datasets and Methods

We use the MovieLens dataset[1] and UCI dataset[2] to testify the performance of our algorithm. We also employ LFW, AR face datasets, and KTH action dataset to show that our method could work for high-dimensional visual recognition task.

We employ SPCA [24], SCCA [20], and SMPCA [17] to perform the comparison. SPCA sparse the transformation matrix by elastic net, while the intermediate matrix used for iteration is updated by SVD of data and the transformation matrix. SCCA performs the dual sparsity processing by elastic net either, which aims to obtain both the sparse coefficient matrices, which is updated alternatively. SMPCA intends to obtain the sparse transformation matrices via distilling main energy of tensors. While STSL performs the dual sparsity for the coefficient transformation in the multi-linear space, in order to calculate the transformation matrices by SVD of the sparse coefficient and high-dimensional data. We also set $\lambda_2 = 0$ in STSL to testify the lasso performance.

---

[1] http://www.grouplens.org/node/73.

[2] http://archive.ics.uci.edu/ml/datasets.html.

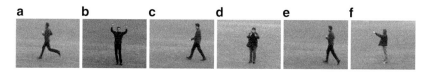

Fig. 3.3 KTH action dataset; the actions (**a**) run, (**b**) wave, (**c**) walk, and (**d**) clap are from Set 1, while (**e**) jog is from Set 2 and (**f**) box is from Set 3

Fig. 3.4 Silhouettes of KTH dataset

Fig. 3.5 The recognition rates of SCCA, SPCA, and TSCC versus the dimensions of KTH action dataset

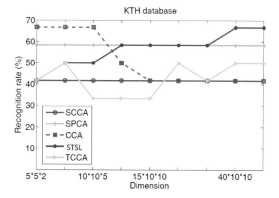

4.2 Experiment on KTH Dataset

We acquire the KTH dataset [16] including six action classes: run (running), box (boxing), wave (handwaving), jog (jogging), walk (walking), and clap (handclapping), as shown in Fig. 3.3. There are four different scenarios: outdoors (s1), outdoors with changing view angles (s2), outdoors where the color of foreground is similar with that of background (s3), and indoors with dim various lighting (s4). We choose 60 samples containing the 6 classes of action from the four scenarios, and fivefold cross validation (CV) is used in this experiment.

Figure 3.4 shows the silhouettes of the box action. Figure 3.5 shows the classification results of different methods under different dimensions. The x-axis displays the tensorial dimensions after projection in STSL, whose product indicates the dimension of the vector-based methods, e.g., $40 \times 10 \times 10 = 4000$. We can see that the accuracy of STSL increases with the larger dimensions, and the number of nonzero loadings (Card) also has to increase correspondingly to preserve more information by selecting the variables. In SCCA and SPCA, the variables are far more than the samples, so we set $\lambda_2 \to \infty$, resulting in the soft-thresholding [23].

We can see that the rate curves of SCCA and SPCA are stable throughout the x-axis with different dimensions. This is because the action silhouettes contain 0/1 value merely, the extracted information from which will not change significantly in SPCA and SCCA. While in CCA, the damaged silhouettes are fragmented, resulting in the more dimensions retained the less efficiency obtained. The curve of TCCA is not stable, due to the projection matrices derived from the damaged silhouettes with less information.

4.3   *Experiment on LFW Face Dataset*

We used the LFW RGB color face dataset [3] containing 1251 faces of 86 individuals with varying expression, view angles, and illuminations. Each face image is normalized to be 32×32 pixels, as shown in Fig. 3.6. We choose ten face images of each individual, and fivefold cross validation (CV) is employed in this experiment.

Figure 3.7 shows the classification rates of different methods under different dimensions. The x-axis displays the tensorial dimensions after projection in STSL, whose product indicates the dimension of the vector-based methods, e.g.,

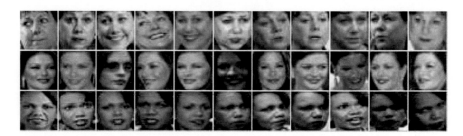

Fig. 3.6 RGB color faces of the LFW dataset

Fig. 3.7 The recognition rates of SCCA, SPCA, and STSL versus the dimensions of LFW face dataset

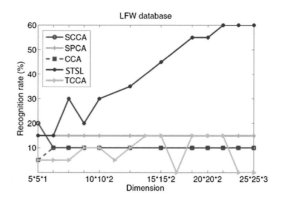

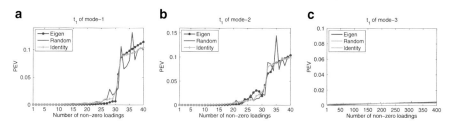

Fig. 3.8 Convex characteristics under projection matrices $\mathbf{T}_n$ on the LFW dataset. (**a**)–(**c**) show the percentage of explained variance (PEV) of sparse loadings t_1 of each mode, respectively

$20 \times 20 \times 2 = 800$. We can see that the accuracy of STSL increases with the larger dimensions, and the number of nonzero loadings (Card) also has to increase correspondingly to preserve more information by selecting the variables. In SCCA and SPCA, the variables are far more than the samples, so we set $\lambda_2 \rightarrow \infty$, resulting in the soft-thresholding [23]. We can see that the rate curves of SCCA, SPCA, and CCA are stable throughout the x-axis with different dimensions, and the corresponding rates are quite low. This is perhaps due to the soft-thresholding rule performed in SCCA and SPCA, which depends on the lasso penalty irrespective of the collinearity caused by the dependency between variables, resulting in large variance of the estimates and devolution. The rather low curve of TCCA may be caused by the projection matrices, which are affected by the changing illusions, view angles, and expressions. Meanwhile CCA may contain more redundant variables which are useless for the classification.

Figure 3.8 shows that STSL is convex irrespective of the original projection matrix $\mathbf{S}_n$ and $\mathbf{T}_n$. We took the dimension $20 \times 20 \times 2$ for illustration. Figure 3.8a–c shows the percentage of explained variance (PEV) by the first dimension of the matrix $\mathbf{T}_n$ against the number of nonzero loadings (Card) in each mode. It can be seen that the PEV of t_1 has the similar trend along the number of nonzero loadings, regardless of the initial $\mathbf{T}_n$.

4.4 Experiment on AR Face Dataset

We use the AR RGB color face dataset [14] of 100 individuals with varying expression, changing illumination, glass wearing, and chin occlusion. Each individual has 26 images captured under four conditions. Each image is normalized to be 32×32 pixels, which are shown in Fig. 3.9. We randomly choose ten people, each with ten face images, respectively. Therefore there are total 100 images (samples) in this experiment. We use fivefold cross validation (CV), which means that there are 80 samples for training while 20 for testing each time. A color face is represented as

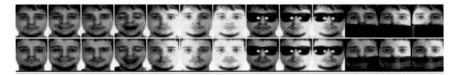

Fig. 3.9 RGB color faces of AR dataset

Fig. 3.10 The recognition rates of CCA, SCCA, SPCA, and TSCC versus the dimensions of AR face dataset

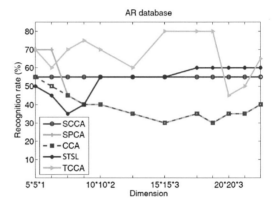

$32 \times 32 \times 3$ size tensor in our algorithm STSL, i.e., the color face is decomposed to be three images, each of which indicates a channel of RGB color space.

The vector-based methods CCA [1], SCCA [12], SPCA [24], and TCCA [7] are used for comparison. In these methods, a color face is represented as a high-dimensional vector size $32 \times 32 \times 3 = 3072$. In CCA, the maximal canonical correlations of two image sets are calculated for classification. SCCA calculates the canonical correlations of the pairwise linear combinations composed of sparse loadings and image vectors. SPCA obtains the sparse principal components as transformation matrix for testing image. TCCA calculates the maximum correlation between two tensors, the combinations of which serve as the projection matrices and perform iteratively. We evaluate the STSL method in terms of accuracy and some other properties, such as shrinkage degree and ridge penalty.

Figure 3.10 shows the classification rates of different methods under various dimensions. The x-axis displays the tensorial dimensions after projection in STSL, whose product indicates the dimension of the vector-based methods, e.g., $20 \times 20 \times 3 = 1200$. We can see that the accuracy of STSL increases with the larger dimensions, and the number of nonzero loadings (Card) also has to increase correspondingly to preserve more information by selecting the variables. In TCCA, the curve of recognition rates is not stable along the various dimensions. Due to that the performance is prone to the projection matrices, which contain the redundancy in the iterative procedure. In SCCA and SPCA, the variables are far more than the samples ($p \gg m$), so we set $\lambda_2 \to \infty$, resulting in the soft-thresholding [23, 24]. We can see that the rate curve of SCCA is stable throughout the x-axis with different dimensions while SPCA and CCA are both decreasing, which because that SCCA

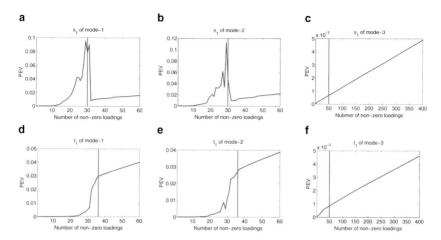

Fig. 3.11 AR dataset: percentage of explained variance (PEV) of the canonical vectors as the function of the number of nonzero loadings. PEV reaches the maximum at the proper nonzero number, and we choose as large the variance as possible under a small number. The *vertical lines* indicate the choice of nonzero number

takes account of the correlations between different samples compared to SPCA and deal well with the occluded or damaged of the original face images by eliminating the redundant pixels different from CCA.

However, in our experiment, we set the parameter Card to fix the number of nonzero variables in the sparse loadings. Because it's much faster and can get higher accuracy than setting s. Figure 3.11 shows the PEV in tensor $\mathcal{X}$ variables, which demonstrates the sparsity performance using the number of nonzero loadings. As displayed in PEV of *mode*-1,2, the PEV obtains its maximum at the proper nonzero number. As few as 50.00–60.00 % of 60 pixels (after *mode-n* projection) can adequately constitute the canonical vectors to bear with such a loss of explained variance (about 0.01 % in v_1), and 12.50 % of 400 pixels with less than 0.005 % of loss in *mode*-3, as PEV of *mode*-3 shows. The x-axis indicates the unfolding dimension after *mode-n* projection as preprocessing. The similar results are also obtained for the other tensor $\mathcal{Y}$ variables.

Table 3.1 shows the classification results of different methods as well as the setting dimension and feature. We select $20 \times 20 \times 2$ as the feature of STSL while 800 in the vector-based methods. We also compare the rates under $\lambda_1 = 0$ and $\lambda_2 = 0$ of STSL, which represent the cases of ridge regression and lasso regression, respectively. Our result is shown in bold value result demonstrates that the elastic net is more suitable as a regression method for recognition in pixel processing compared to the ridge and the lasso.

Table 3.1 The recognition rates (percent) of CCA, SCCA, SPCA, and STSL on a subset of the AR dataset and the corresponding dimensions and feature

Method	CCA	SCCA	SPCA	TCCA	STSL [Ours]	STSL ($\lambda_1 = 0$)	STSL ($\lambda_2 = 0$)
Accuracy	30	55	35	45	**60**	15	50
Dimension	3072			$32 \times 32 \times 3$			
Feature	800			$20 \times 20 \times 3$			

Table 3.2 Datasets

Dataset	Objects	Variate	Size	Dimensions	Class
MovieLens	100,000	4	7,121	$5 \times 2 \times 2$	–
Breast-w	683	9	683	3×3	2
Vehicle	846	18	846	$3 \times 3 \times 2$	4
Heart-statlog	270	13	270	$3 \times 2 \times 2$	2
Letter	20,000	16	20,000	$4 \times 2 \times 2$	26
Sonar	208	60	208	$6 \times 2 \times 5$	2
Ionosphere	351	34	351	17×2	2
Clean1	476	166	475	$20 \times 2 \times 4$	2
Diabetes	768	8	768	$2 \times 2 \times 2$	2

4.5 MovieLens and UCI Datasets

The MovieLens dataset has five subsets, each of which contains 100,000 items, including three variates: movie-ID, scores, and time. We employ the first two variates, with ten items of each user ID as a tensor, so there are more than 1500 tensors excluding the users less than ten items to be used in each subset. We also extracted eight real datasets from the UCI dataset for accuracy, run time, and other performance. They are Breast-w, Vehicle, Heart-statlog, Letter, Sonar, Ionosphere, Clean1, and Diabetes, respectively. We transformed the number of variates into the tensor dimensions, and the details of the datasets are shown in Table 3.2.

4.5.1 RMSE and Accuracy

Figure 3.12 shows the area under curve (AUC) value of different methods. The upper left of the curve indicates its superiority, which means more accuracy with less error. It shows that our algorithm performs better than SMPCA. SCCA and SPCA runs litter better in some cases, such as Figs. 3.12b, c; however, TFDS performs competitive in most cases with larger AUC. Table 3.3 shows the RMSE of different datasets. The RMSE and variance of STSL is less than those of others. Specifically, STSL performs better with the elastic net constraint in most cases compared with lasso. Table 3.4 demonstrates the accuracy under the original dimensions. The

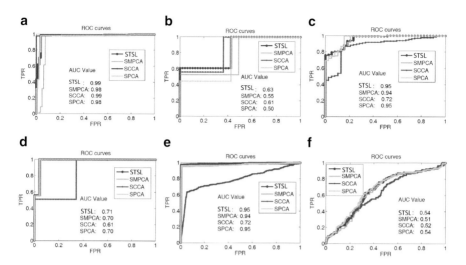

Fig. 3.12 ROC curves on different datasets. (**a**) Breast-w. (**b**) Sonar. (**c**) Ionosphere. (**d**) Clean1. (**e**) Letter. (**f**) Diabetes

accuracy of STSL is comparative with those of others. Table 3.5 shows the run time w.r.t. of different datasets, and the time complexity of STSL is obviously lower than that of SMPCA. The time costing also decreases when lasso is replaced by elastic net, which alleviates the sparse degree. All the tensorial solutions have reasonable time efficiency compared to the vectorial methods, which indicates that the tensorial representation can be used for high-dimensional data processing, for saving time and memory.

4.5.2 Explained Variance Analysis

The explained variance [12] is used to compare the efficiency of sparsity. More sparsity with larger explained variance (EVar) indicates that we eliminate the redundance while keeping the significant variates, which is meaningful for processing high-dimensional data and large dataset.

"Given a N-order tensor y with I_n ($n = 1, \ldots, N$), dimension in each mode, the corresponding *mode-n* unfold matrix is denoted as $Y^{(n)}$", which is rewritten as Y for simplicity, I_n is denoted as I as well.

"Initialize two *mode-n* transformation matrices T_n and S_n, which are denoted as T and S for simpicity, here we only consider the first two projection vectors T_1, S_1, T_2 and S_2".

$$\text{EVar}(\mathbf{Y}|\mathbf{T}_1) = \frac{\text{tr}[\text{Var}(\text{E}(\mathbf{Y}|\mathbf{T}_1))]}{\text{tr}[\text{Var}(\mathbf{Y})]} = \frac{1}{I}\rho(\mathbf{T}_1, \mathbf{Y})\rho(\mathbf{Y}, \mathbf{T}_1), \tag{3.14}$$

where ρ is the correlation between two variates.

Table 3.3 RMSE w.r.t. Datasets

Dataset (Size, Variate, Dimension)				SPCA	SCCA	SMPCA	STSL ($\lambda_2 = 0$)	STSL
MovieLens (u_1)	1324	20	$5 \times 2 \times 2$	25.7737 ± 123.5791	26.3439 ± 136.3012	2.2155 ± 0.9061	0.5494 ± 0.6555	0.3968 ± 0.0288
MovieLens (u_2)	1412	20	$5 \times 2 \times 2$	27.8994 ± 117.5267	26.0017 ± 102.9568	2.6363 ± 1.0373	0.7856 ± 0.0919	0.8008 ± 0.0957
MovieLens (u_3)	1468	20	$5 \times 2 \times 2$	29.5763 ± 117.6812	28.4270 ± 111.1187	2.7553 ± 1.0134	0.7356 ± 0.0721	0.3939 ± 0.0158
MovieLens (u_4)	1457	20	$5 \times 2 \times 2$	30.0588 ± 115.5300	30.0817 ± 117.3369	2.7497 ± 0.9660	0.3681 ± 0.0173	0.6592 ± 0.0555
MovieLens (u_5)	1460	20	$5 \times 2 \times 2$	31.1069 ± 128.0725	30.3609 ± 122.8663	2.7675 ± 1.0116	0.8798 ± 0.1020	0.3109 ± 0.0128
Breast-w	683	9	3×3	2.6017 ± 0.9543	2.1062 ± 0.6246	2.8486 ± 1.3099	1.6332 ± 0.3052	1.5876 ± 0.2895
Vehicle	846	18	$3 \times 3 \times 2$	29.3509 ± 13.9579	29.4552 ± 11.3440	17.5151 ± 2.1563	6.6033 ± 0.2356	7.2807 ± 0.4011
Heart	270	12	$3 \times 2 \times 2$	21.1430 ± 1.8029	22.1156 ± 1.5891	6.0479 ± 0.1373	0.9423 ± 0.0033	0.8698 ± 0.0028
Letter	20,000	16	$4 \times 2 \times 2$	6.1186 ± 0.1739	5.5594 ± 0.0776	2.2603 ± 0.0533	0.3808 ± 0.0005	0.3986 ± 0.0007
Sonar	208	60	$6 \times 2 \times 5$	1.7540 ± 0.0109	2.0712 ± 0.0156	1.6231 ± 0.0099	1.4764 ± 0.0177	0.5118 ± 0.0027

Table 3.4 Recognition rate (%) w.r.t. Datasets

Dataset (Size, Dimension)			SPCA	SCCA	SMPCA	STSL ($\lambda_2 = 0$)	STSL
Breast-w	683	3×3	97.67	98.25	96.50	97.67	98.25
Ionosphere	351	17×2	93.14	64.57	92.00	92.00	92.57
Heart	270	$3 \times 2 \times 2$	55.56	56.30	55.56	57.04	57.78
Sonar	208	$6 \times 2 \times 5$	50.00	59.26	51.85	50.93	59.26
Clean1	475	$20 \times 2 \times 4$	67.76	55.51	67.35	66.94	68.16
Diabetes	768	$2 \times 2 \times 2$	66.67	56.77	65.36	64.84	65.36

Table 3.5 Runtime (sec.) w.r.t. Datasets

Dataset (Size)		SPCA	SCCA	SMPCA	STSL ($\lambda_2 = 0$)	STSL
Breast-w	683	2.48	0.08	0.76	0.88	0.16
Ionosphere	351	7.43	0.25	0.72	1.03	0.92
Heart	270	1.66	0.07	0.66	0.21	0.18
Sonar	208	9.49	0.37	2.79	0.52	1.50
Clean1	475	64.32	2.29	15.89	1.64	3.46
Diabetes	768	2.20	0.37	0.98	0.20	0.19

The *mode*-2 explained that variance is computed by getting rid of the correlations between *mode*-1 and *mode*-2 sample, as

$$\mathrm{EVar}(\mathbf{Y}|\mathbf{T}_2) = \frac{\mathrm{tr}[\mathrm{Var}(\mathrm{E}(\mathbf{Y} - \mathrm{E}(\mathbf{Y}|\mathbf{S}_1, \mathbf{T}_1)|\mathbf{T}_2))]}{\mathrm{tr}[\mathrm{Var}(\mathbf{Y})]}$$

$$= \frac{1}{I}\rho\left(\mathbf{Y}', \mathbf{T}_2\right)\rho\left(\mathbf{T}_2, \mathbf{Y}'\right), \tag{3.15}$$

where $\mathbf{Y}' = \mathbf{Y} - \frac{1}{I}\rho(\mathbf{T}_1, \mathbf{Y})\rho(\mathbf{Y}, \mathbf{S}_1)$.

In this part, we compute the explained variance of $\mathbf{Y}$ in the heart-statlog dataset, the same as $\mathbf{X}$. The variates selected and the explained variance of all the methods are shown in Table 3.6, from which we can see that the EVar of STSL is competitive with others in both the first and second dimensions of data, with the EVar of the former larger than the latter. Especially, the number of selected variates of STSL is obviously less than that of SMPCA, which indicates the superior performance of STSL with larger EVar and less variates.

4.5.3 Computational Cost

Given m N-order tensor $\mathcal{A} \in \mathbb{R}^{I_1 \times \cdots \times I_N}$. For simplicity, we suppose that $I_1 = \cdots = I_N = (\prod_{n=1}^{N} I_n)^{(1/n)} = I$. The main complexity cost steps are computing

Table 3.6 Variance and selected variates on heart-statlog dataset

Variate	1st dim.					2nd dim.				
	SPCA	SCCA	SMPCA	STSL ($\lambda_2 = 0$)	STSL	SPCA	SCCA	SMPCA	STSL ($\lambda_2 = 0$)	STSL
Age	0	0	0.0199	0	0	0	0	0.1591	0	0
Sex	0	0	0.0515	0	0	0	0	-0.0014	0	0
Chest pain	0	0	0	0	0	0	0	0	0	0
Blood pressure	0.5552	0.0202	-0.0817	0	0	0	0	-0.6532	0	0
Cholesterol	0.8317	0.9950	-0.2113	0	0	0.0056	-0.0128	0.0057	0	0
Blood sugar	0	0	0	0	0	0	0	0	0	0
Electro-result	0	0	0.0005	2.4248×10^{-15}	-1.0000×10^{-12}	0	0	0.0038	0	-1.0000×10^{-12}
Heart rate	0	0	0.0012	7.6910×10^{-17}	0	1.0000	-0.9347	-3.3212×10^{-5}	0	0
Exercise (angina)	0	0	0	0	3.2000×10^{-11}	0	0	0	0	2.2000×10^{-11}
Oldpeak	0	0	-0.0019	0	0	0	0	-0.0156	-5.3213×10^{-16}	0
ST segment	0	0	-0.0050	0	0	0	0	1.3637×10^{-4}	-1.6187×10^{-17}	0
Major vessels	0	0	0	0	0	0	0	0	0	0
EVar of Y										
Mode-1	0.5052	1.2028×10^{-6}	5.8795	6.4570	9.0668	0.5321	1.3822×10^{-6}	0.5017	2.6519	2.7470
Mode-2			0.6876	1.6547	1.4510			29.0643	0.2836	0.2180
Mode-3			1.3222	1.4141	1.2992			0.1302	0.1421	0.1106

the covariance matrix and constraint condition. Therefore the cost time for STSL is $O\left(N \cdot I^2 \cdot m + 2N \cdot I\right)$, while for SMPCA is $O\left(N \cdot I \cdot m^2 + 2N \cdot I\right)$ $(m \gg I)$, and for SPCA is $O\left(I^{2N} \cdot m + 2I^N\right)$. We can see that the computational cost of STSL is far less than that of SMPCA and SPCA.

4.5.4   Discussion

The RMSE of vectorial methods SPCA and SCCA are quite high; this is because each data has ten items, each with movie-ID and a score. SPCA and SCCA calculate the RMSE of all the ten items, while the tensorial solutions only compute the RMSE of the score, which is the second mode of a tensor. The different processing means that we can deal with the data with meaningful section, as long as it is in tensorial representation. Because of the different computation of covariance matrix, STSL has less time complexity than SMPCA. Compared with SPCA, the less dimensions of variates in tensorial subspace of our method results in the less time consuming. ROC curves illustrate that the accuracy of STSL is competitive with other methods in different datasets. And the SPCA also performs good in many cases. The explained variance of samples indicates that we eliminate the redundance, while selecting the significant variates for data representation. The STSL is competitive with other sparse methods according to the explained variance in tensorial subspace. STSL does not resemble traditional canonical correlation analysis, because it performs SVD on the product of data covariance and sparse coefficients alternatively, instead of simply linear combination, to obtain the invariant structure of the data. By the tensorial representation of the original data, we can deal with the high-dimensional data not only to save time and memory but also to process the meaningful part we need and obtain the solution pertinently.

5   Summary

We propose a novel STSL model, for dimensional reduction. The tensorial representation of data has less dimensions then a vector, which is suitable for processing high-dimensional data; and the different mode can indicate various meaning of data. For high-dimensional data, like human action video with spatiotemporal information, STSL can select the key frames to reduce noise and redundancy, meanwhile learning a subspace for dimensional reduction. STSL has lower time complexity and can save memory at the same time. The accuracy of STSL is competitive with the art-of-the-state methods. The explained variance of sample computed by STSL is competitive with other vectorial and tensorial methods on the heart-statlog dataset and common face datasets, because STSL obtains larger explained variance with less selected variates. STSL employs the elastic net constraint instead of lasso, for two reasons: firstly, ridge regression is used for decorrelation, to avoid singular solution, and, secondly, the rescaling factor is taken to eliminate double sparsity. The accuracy of different constraints shows the optimization.

The direction of this work can add discriminant analysis in the updating step, by performing SVD on the product of interclass, intra-class covariance and sparse coefficient vector, to update the transformation matrices.

References

1. Bach, F., Jordan, M.: A probabilistic interpretation of canonical correlation analysis. TR 688, Department of Statistics, University of California, Berkeley (2005)
2. Hardoon, D., Shawe-Taylor, J.: Sparse canonical correlation analysis. Mach. Learn. **83**, 331–353 (2011)
3. Huang, G., Mattar, M., Berg, T., Learned-Miller, E., et al.: Labeled faces in the wild: a database for studying face recognition in unconstrained environments. In: Workshop on Faces in 'Real-Life' Images: Detection, Alignment, and Recognition (2008)
4. Jia, C., Kong, Y., Ding, Z., Fu, Y.: Latent tensor transfer learning for RGB-D action recognition. In: Proceedings of the ACM International Conference on Multimedia, pp. 87–96. ACM (2014)
5. Jia, C., Zhong, G., Fu, Y.: Low-rank tensor learning with discriminant analysis for action classification and image recovery. In: Twenty-Eighth AAAI Conference on Artificial Intelligence (2014)
6. Kim, T., Cipolla, R.: Canonical correlation analysis of video volume tensors for action categorization and detection. IEEE Trans. Pattern Anal. Mach. Intell. **31**(8), 1415–1428 (2008)
7. Kim, T., Wong, S., Cipolla, R.: Tensor canonical correlation analysis for action classification. In: Conference on Computer Vision and Pattern Recognition, pp. 1–8. IEEE (2007)
8. Kong, Y., Fu, Y.: Bilinear heterogeneous information machine for RGB-D action recognition. In: Proceedings of the IEEE Conference on Computer Vision and Pattern Recognition, pp. 1054–1062 (2015)
9. Kong, Y., Jia, Y., Fu, Y.: Interactive phrases: semantic descriptions for human interaction recognition. IEEE Trans. Pattern Anal. Mach. Intell. **36**(9), 1775–1788 (2014)
10. Li, K., Hu, J., Fu, Y.: Modeling complex temporal composition of actionlets for activity prediction. In: Computer Vision–ECCV 2012, pp. 286–299. Springer (2012)
11. Lu, H., Plataniotis, K., Venetsanopoulos, A.: MPCA: multilinear principal component analysis of tensor objects. IEEE Trans. Neural Netw. **19**(1), 18–39 (2008)
12. Lykou, A., Whittaker, J.: Sparse CCA using a lasso with positivity constraints. Comput. Stat. Data Anal. **54**(12), 3144–3157 (2010)
13. Mardia, K.V., Kent, J.T., Bibby, J.M.: Multivariate Analysis. Academic, London (1980)
14. Martinez, A., Benavente, R.: The AR face database. CVC Technical Report 24, University of Purdue (1998)
15. Parkhomenko, E., Tritchler, D., Beyene, J.: Sparse canonical correlation analysis with application to genomic data integration. Stat. Appl. Genet. Mol. Biol. **8**(1), 1 (2009)
16. Schuldt, C., Laptev, I., Caputo, B.: Recognizing human actions: a local SVM approach. In: International Conference on Pattern Recognition, vol. 3, pp. 32–36. IEEE (2004)
17. Shashua, A., Levin, A.: Linear image coding for regression and classification using the tensor-rank principle. In: IEEE Conference on Computer Vision and Pattern Recognition, vol. 1, pp. I–42. (2001)
18. Sun, C., Junejo, I., Tappen, M., Foroosh, H.: Exploring sparseness and self-similarity for action recognition. IEEE Trans. Image Process. **24**(8), 2488-2501 (2015)
19. Tao, D., Li, X., Wu, X., Maybank, S.: General tensor discriminant analysis and gabor features for gait recognition. IEEE Trans. Pattern Anal. Mach. Intell. **29**, 1700–1715 (2007)

20. Waaijenborg, S., Verselewel de Witt Hamer, P., Zwinderman, A.: Quantifying the association between gene expressions and DNA-markers by penalized canonical correlation analysis. Stat. Appl. Genet. Mol. Biol. **7**(1), 1–29 (2008)
21. Yan, S., Xu, D., Yang, Q., Zhang, L., Tang, X., Zhang, H.: Discriminant analysis with tensor representation. In: Conference on Computer Vision and Pattern Recognition, vol. 1, pp. 526–532 (2005)
22. Zhang, J., Han, Y., Jiang, J.: Tucker decomposition-based tensor learning for human action recognition. Multimedia Syst., 1–11 (2015). ISSN: 0942-4962
23. Zou, H., Hastie, T.: Regularization and variable selection via the elastic net. J. R. Stat. Soc. Ser. B (Stat Methodol.) **67**(2), 301–320 (2005)
24. Zou, H., Hastie, T., Tibshirani, R.: Sparse principal component analysis. J. Comput. Graph. Stat. **15**(2), 265–286 (2006)

Chapter 4
Multimodal Action Recognition

Chengcheng Jia, Wei Pang, and Yun Fu

1 Introduction

1.1 Tensor for Action Recognition

For many machine learning and pattern recognition tasks, most of the data are high dimensional, such as human action videos. Vectorization of action video is one common approach to dealing with the high dimensionality issue. However, the vectorization process requires a huge amount of memory, and it is also very time consuming. Recently, tensor decomposition analysis (TDA) [1, 9, 29] has been successfully applied to various high-dimensional recognition related problems, such as action recognition [7, 17, 20], face recognition [3, 8, 27], and gait recognition [6, 15, 28]. TDA represents high-dimensional data as a multi-fold (mode) tensor,

---

© {Chengcheng Jia, Wei Pang, and Yun Fu | Springer}, {2015}. This is a minor revision of the work published in {Proceedings of the Computer Vision - ECCV 2014 Workshops, pp. 818–833. Springer Volume 8925 of the series Lecture Notes in Computer Science, 2015}, http://dx.doi.org/10.1007/978-3-319-16178-5-57.

C. Jia (✉)
Department of Electrical and Computer Engineering, Northeastern University,
360 Huntington Avenue, Boston, MA 02115, USA
e-mail: jia.ch@husky.neu.edu

W. Pang
School of Natural and Computing Sciences, University of Aberdeen, Aberdeen, UK
e-mail: pang.wei@abdn.ac.uk

Y. Fu
Department of Electrical and Computer Engineering and College of Computer and Information Science (Affiliated), Northeastern University, 360 Huntington Avenue, Boston, MA 02115, USA
e-mail: yunfu@ece.neu.edu

© Springer International Publishing Switzerland 2016
Y. Fu (ed.), *Human Activity Recognition and Prediction*,
DOI 10.1007/978-3-319-27004-3_4

which could keep the original structure of data instead of vectorization. TDA is essentially the extension of vector or matrix analysis to higher-order tensor analysis. In the tensor subspace, the discriminant projection matrix of each mode [14] is calculated alternately by fixing the other modes. Some TDA methods integrate Fisher criteria on each mode, such as general tensor discriminant analysis (GTDA) while some TDA methods find out the correlation of data from each mode, such as multi-linear discriminant canonical correlation analysis (MDCC [11]).

Tensor discriminant analysis employed Fisher criteria on each mode of tensor, e.g., GTDA [22] for gait recognition and DATER [26] for face recognition. GTDA and DATER not only preserve the original high-dimensional structure of data, but also avoid the "curse of dimensionality" caused by small datasets. Consequently, they achieved good results on gait recognition and face recognition. However, most tensor-based methods have been performed directly on samples [6, 8], which may decrease the accuracy in situations where image occlusion or damage exists.

Considering the variations of an object due to different angles and illuminated conditions, canonical correlation analysis (CCA) [5, 10, 19, 24] is often used for computing the similarity of two datasets to overcome this problem. Kim et al. [13] proposed the tensor canonical correlation analysis (TCCA) by calculating the correlation of two tensor samples for action detection and recognition. TCCA improves the accuracy compared with CCA and it is more time efficient for action detection. In our previous work [11], a CCA-based discriminant analysis in multi-linear subspace (MDCC) is proposed for action recognition. In that work, we took an action image sequence as a *3-order* tensor, and calculated the discriminant projection matrices using canonical correlations between any pairwise of tensors. MDCC with discriminant information obtained better accuracy than TCCA.

However, all the above mentioned methods perform Fisher criteria [2] or CCA on an action tensor individually, irrespective of the different realistic meanings of each mode, e.g., the pixel character of the image mode and the correlation of the sequence mode. Since an action image sequence can be seen as a *3-order* tensor, of which *mode-*1 and *mode-*2 represent the action image information and *mode-*3 describes the temporal information [13], in a similar way, an RGB object image can also be represented as a *3-order* tensor, of which the first and second modes stand for an image, while the third mode indicates lateral illumination or RGB color. In general case, Fisher criteria can be used to extract image features for classification. In addition, we also note that the correlations among the images can reflect the temporal ordering of an action or among various RGB colors of an object caused by multi-view and illumination, in this sense, by considering the correlations among images we can improve the performance of the subsequent recognition task [4, 18, 25].

1.2 Our Contribution

In this chapter we explore a mode alignment method with different criteria which employs Fisher criteria on the image modes and correlation analysis on the sequence mode, and such mode alignment method is termed mode-driven discriminant

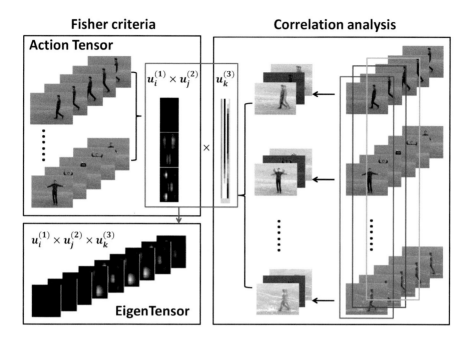

Fig. 4.1 The framework of MDA. The *left* is *mode*-1, 2 aligned by Fisher criteria, while the *right* is frame alignment along *mode*-3, containing I_3 subsets (each is shown in *red, blue, green boxes*). We update the action sequence by using the correlation between three adjacent subsets, then we perform the Fisher criteria on *mode*-3

analysis (MDA) in this chapter [12]. The reason for using different criteria in different modes is that different modes have different visual distributions, i.e., the discrete distribution means the image subspaces in the first $N - 1$ modes, while the continuous distribution means the sequential patterns in the Nth mode. The proposed framework is shown in Fig. 4.1.

In this framework, there are m samples in the whole dataset, and we extract the *mode*-1,2 spatial information of an action by discriminant criterion. Suppose an action is composed of I_3 frames, we re-organized the dataset to be I_3 subsets, each of which contains m frames. We call this **frame-to-frame** reconstruction of an action sequence, i.e., we update the *mode*-3 tensor by calculating the correlation of their adjacent frames, then, we use the correlation analysis on the updated *mode*-3. We propose to make the best use of different representation of data, i.e., for the spatial pattern, we extract the features directly from raw data represented by pixels, while for the sequential pattern, we aim to extract the features by finding the intrinsic correlation between the adjacent frames. This idea can be generalized to other representation of data with various factors, such as pose, illumination, multi-view, expression, color for face recognition, and multi-view, subject diversity, depth for action recognition. We let the iterative learning subject to a plausible convergence condition. The projection is both discriminant and canonical correlated, in contrast to traditional approaches [8, 13]. Hence we achieve good accuracy on action and object recognition even with a simple nearest neighbor classifier (NNC). We summarize the advantages of the proposed MDA as follows.

1. MDA employs a mode alignment method with specific discriminant criteria on different modes of one tensor.
2. MDA considers the correlation between adjacent frames in the temporal sequence or RGB pattern by a **frame-to-frame** reconstruction approach.
3. MDA converges as proved in Sect. 2.2 and discussed in Sect. 3.

The rest of this chapter is organized as follows: Sect. 2.1 introduces the fundamentals of tensor and canonical correlation analysis. Section 2.2 describes in detail the MDA algorithm for dimensional reduction. Experiments on action recognition and object recognition using the proposed method and other compared methods are reported in Sect. 3. Finally, conclusions are given in Sect. 4.

2 Proposed Method

In this section, we first introduce some fundamentals about tensor decomposition; then we overview the framework; third we briefly introduce the Fisher analysis on image mode; fourth, we give the details of correlation analysis of time series on the temporal mode.

2.1 Fundamentals

In this part, we introduce some fundamentals about tensor decomposition and CCA. A multi-dimensional array $\mathcal{A} \in \mathbb{R}^{I_1 \times \cdots \times I_n \times \cdots \times I_N}$ is called an *N-order* tensor, where I_n is the dimension of *mode-n*. $\mathcal{A}_{i_1 \ldots i_N}$ is an element of $\mathcal{A}$. A tensor can usually be unfolded, which means one mode is fixed, while other modes are stretched to form a huge matrix $\mathbf{A}^{(n)}$ [14].

Definition 1 (Tucker Decomposition [14]). $\mathcal{A} \in \mathbb{R}^{I_1 \times \cdots \times I_N}$ is an *N-order* tensor and $\mathbf{U}_n \in \mathbb{R}^{I_n \times J_n} (1 \leqslant n \leqslant N)$ is used for decomposition, we have the following equation:

$$\mathcal{S} = \mathcal{A} \times_1 \mathbf{U}_1^{\mathrm{T}} \times_2 \mathbf{U}_2^{\mathrm{T}} \ldots \times_n \mathbf{U}_n^{\mathrm{T}} \ldots \times_N \mathbf{U}_N^{\mathrm{T}}. \tag{4.1}$$

The above equation indicates the procedure of dimension reduction of a tensor, and $\mathcal{S}$ is called the core tensor, $\times_n$ is *mode-n* product.

Definition 2 (Canonical Correlation [19]). Given two vectors $x \in \mathbb{R}^m$, $y \in \mathbb{R}^n$, and two coefficient vectors $\alpha \in \mathbb{R}^m$, $\beta \in \mathbb{R}^n$, the correlation of x and y is calculated by

$$\rho(u, v) = \underset{u,v}{\arg\max} \ \frac{\text{Cov}(u, v)}{\sqrt{\text{Var}(u)} \sqrt{\text{Var}(v)}}, \tag{4.2}$$

subject to variance $\text{Var}(u) = 1$ and $\text{Var}(v) = 1$, where $u = \alpha^T x$ and $v = \beta^T y$, $\rho(u, v)$ and $\text{Cov}(u, v)$ are the canonical correlation and covariance matrix, respectively.

2.2 Frame Workflow

In this subsection, we propose MDA which uses Fisher criteria to transform the image information and uses CCA to transform the temporal information. It is noted that the discriminant analysis on an action image sequence can extract more effective information which reflects the properties of images [22, 26]. Also for temporal space, correlations among the images can reflect the temporal ordering of an action in subsequent recognition methods [4]. In our tensor based framework, we first introduce the tensor Fisher analysis on image modes, then explain CCA on the temporal mode, respectively.

In this subsection, we introduce the discriminant criteria of MDA for both image and temporal modes. An action sample is represented as a *3-order* tensor. The *mode-1,2* represent spatial dimension and the *mode-3* represents temporal information. Suppose there are m tensor samples in C classes, and $\mathcal{A} \in \mathbb{R}^{I_1 \times I_2 \times I_3}$ is a tensor sample. Our goal is to find the transformed matrices $\mathbf{U}_n \in \mathbb{R}^{I_n \times J_n}$ $(1 \leqslant n \leqslant 3)$ for projection in tensor space to achieve dimension reduction. $\mathbf{U}_n$ is calculated alternately by fixing the other modes. $\mathbf{U}_n$ is defined with the *mode-n* discriminant function $\boldsymbol{F_n}$ as shown below

$$\begin{aligned} \mathbf{U}_n &= \underset{\mathbf{U}_n}{\arg\max} \ \boldsymbol{F_n} \\ &= \underset{\mathbf{U}_n}{\arg\max} \ tr\left(\mathbf{U}_n^T(\mathbf{S}_b^{(n)} - \alpha\mathbf{S}_w^{(n)})\mathbf{U}_n\right), \end{aligned} \tag{4.3}$$

where tr is the sum of diagonal elements of matrix $\left(\mathbf{U}_n^T(\mathbf{S}_b^{(n)} - \alpha\mathbf{S}_w^{(n)})\mathbf{U}_n\right)$, and α is a tuning parameter.

2.3 Fisher Analysis on Image Mode

In this subsection, we introduce the Fisher criteria on the image mode. Given a tensor set with m samples, each of which $\mathcal{A}_{ij} \in \mathbb{R}^{I_1 \times \cdots \times I_N}$ indicates the jth sample belonging the ith class.

In the case of $1 \leq n \leq N - 1$, $\mathbf{S}_b^{(n)}$ is the inter-class scatter matrix calculated by

$$\mathbf{S}_b^{(n)} = \frac{1}{m} \sum_{i=1}^{C} m_i \left(\bar{\mathbf{A}}_i^{(n)} - \bar{\mathbf{A}}^{(n)}\right)\left(\bar{\mathbf{A}}_i^{(n)} - \bar{\mathbf{A}}^{(n)}\right)^T, \tag{4.4}$$

and $\mathbf{S}_w^{(n)}$ is the intra-class scatter matrix calculated as follows:

$$\mathbf{S}_w^{(n)} = \frac{1}{m} \sum_{i=1}^{C} \sum_{j=1}^{m_i} \left(\mathbf{A}_{ij}^{(n)} - \bar{\mathbf{A}}_i^{(n)}\right)\left(\mathbf{A}_{ij}^{(n)} - \bar{\mathbf{A}}_i^{(n)}\right)^T, \tag{4.5}$$

where m_i is the number of the ith class; $\bar{\mathbf{A}}^{(n)}$ is the mean of training samples of *mode-n*; $\bar{\mathbf{A}}_i^{(n)}$ is the mean of ith class (C_i) of *mode-n*; and $\mathbf{A}_{ij}^{(n)}$ is the jth sample of C_i. In the end, $\mathbf{U}_n$ is composed of the eigenvectors corresponding to the largest J_n eigenvalues of $(\mathbf{S}_b^{(n)} - \alpha \mathbf{S}_w^{(n)})$.

In Fisher criteria, the discriminant function $\boldsymbol{F}_n$ is calculated by the original data. While in discriminant CCA, $\boldsymbol{F}_n$ is calculated by the canonical tensor, which is detailed in Sect. 2.4.

2.4 Correlation of Time Series

In this part, we introduce the correlation criteria on the image mode. An action sequence is actually a time series of frames, ranking orderly. We aim to make better use of the correlations between two adjacent frames, which can reflect the variation of an action as time elapses. So, we choose to update the dataset by exploring the correlations between the time series.

In the tensor set, we re-organized the new subsets by collecting the kth ($1 \leq k \leq I_N$) frame of the samples, so there are I_N subsets in the dataset. Our strategy is as follows: first update the new subsets, then re-organize these subsets to form a new dataset, and finally perform the discriminant criterion.

We consider a sample $\mathcal{A}_{ij}$ as an action sequence $\{f_1, \ldots, f_k, \ldots, f_{I_N}\}$, and we organized a new subset $\mathcal{A}_k$ by extracting each kth frame from all the samples, then we calculate the correlation between $(k-1)$th, kth, and $(k+1)$th subsets, which is indicated as $\rho_{k-1,k,k+1}$, as shown in Fig. 4.2. Our goal is to decompose the whole tensor set into I_N subsets, then update the whole tensor sequence by the correlations of the subsets. So, the new dataset has explored the correlation in the time series, which is consistent with the real situation and can describe the time series more precisely. The procedure is detailed as follows.

Fig. 4.2 Illustration of
correlation analysis on the
time series subspace

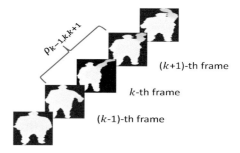

$(k+1)$-th frame

k-th frame

$(k-1)$-th frame

In the case of mode $n = N$ (here $N = 3$), we organized the kth subset by
collecting the kth frames of the dataset as $A_k \in \mathbb{R}^{(I_1 I_2) \times m}$. The singular value
decomposition (SVD) [14] is performed on $\mathbf{A}_k \mathbf{A}_k^{\mathrm{T}}$, as shown below:

$$\mathbf{A}_k \mathbf{A}_k^{\mathrm{T}} = \mathbf{P} \Lambda \mathbf{P}^{\mathrm{T}}, \tag{4.6}$$

where Λ is the diagonal matrix and $\mathbf{P} \in \mathbb{R}^{(I_1 I_2) \times J_N}$ is the matrix composed by the
eigenvectors of the J_N largest eigenvalues, and J_N is the subspace dimension. We
perform CCA on $(k-1)$th, kth and $(k+1)$th sets as follows:

$$\mathbf{P}_{k-1}^{\mathrm{T}} \mathbf{P}_k = \mathbf{Q}_{k-1,k} \Lambda \mathbf{Q}_{k,k-1}^{\mathrm{T}}, \quad \mathbf{P}_k^{\mathrm{T}} \mathbf{P}_{k+1} = \mathbf{Q}_{k,k+1} \Lambda \mathbf{Q}_{k+1,k}^{\mathrm{T}}, \tag{4.7}$$

where Λ is the diagonal matrix and $\mathbf{Q}_{k-1,k}$, $\mathbf{Q}_{k,k+1} \in \mathbb{R}^{J_N \times J_N}$ are orthogonal rotation
matrices. The kth subset $\mathbf{P}_k$ is updated as follows:

$$\mathbf{P}_{k-1,k,k+1} \leftarrow \mathrm{SVD}(\mathbf{P}_k \mathbf{Q}_{k,k+1}^{\mathrm{T}} \mathbf{P}_k^{\mathrm{T}} \mathbf{P}_k \mathbf{Q}_{k,k-1}). \tag{4.8}$$

Then, we update the dataset by the new subsets $\mathbf{A}_k = \{\mathbf{P}_{k-1,k,k+1} | k = 1, \ldots, I_N\}$.
Finally, we perform discriminant analysis on the *mode-N* samples by Eqs. (4.4) and
(4.5).

MDA is different from TCCA [13] when calculating the transformation
matrix (TM). In TCCA, all the TMs are composed of the coefficients vectors
of data by CCA, and there is no discriminant analysis, while in MDA, all the TMs
are calculated by performing discriminant analysis on data. What's more, TCCA
calculates the TM by pairwise tensors, while MDA calculates the inter-class and
intra-class scatter matrices by making use of the correlation of multiple datasets.
The iterative procedure of the proposed method is detailed in Algorithm 1.

Definition 3 (Similarity). The similarity of two tensors, which is essentially the
mode-N correlation between two tensors, is defined as follows:

$$S = (\mathbf{P}_i^{(N)} \mathbf{Q}_{ik}^{(N)})^T (\mathbf{P}_k^{(N)} \mathbf{Q}_{ki}^{(N)}), \tag{4.9}$$

where i, k indicate a pair of tensors. S is called the similarity matrix.

Algorithm 1 Mode-Driven Discriminant Analysis (MDA)

INPUT: m N-order tensors $\Gamma_m = \{\mathcal{A}_i^{C_i}\}$, with label $\{C_i\}$, $(1 \leq i \leq m)$, the tuning parameter α, and the maximum iterations t_{max}.
OUTPUT: Updated $U_n^{(t)}$, $1 \leq n \leq$ N.

1: Initialize $\mathbf{U}_n$ by eigen-decomp of $\mathcal{A}_i^{C_i}$, $1 \leq i \leq m$.
2: **for** $t = 1$ to t_{max} **do**
3:     **for** $n = 1$ to N **do**
4:         $\mathcal{A}_i^{C_i} \leftarrow \mathcal{A}_i^{C_i} \times_1 (\mathbf{U}_1^{(t-1)})^{\mathrm{T}} \cdots \times_{n-1} (\mathbf{U}_{n-1}^{(t-1)})^{\mathrm{T}} \times_{n+1} (\mathbf{U}_{n+1}^{(t-1)})^{\mathrm{T}} \cdots \times_{\mathrm{N}} (\mathbf{U}_{\mathrm{N}}^{(t-1)})^{\mathrm{T}}.$
5:         **if** $n = $ N **then**
6:             Update $\mathcal{A}_i$ via Eqs. (4.6), (4.7), and (4.8).
7:         **end if**
8:         Calculate $\mathbf{S}_b^{(n)}$ and $\mathbf{S}_w^{(n)}$ by Eqs. (4.4) and (4.5).
9:         Update $\mathbf{U}_n^{(t)}$ by eig-decomp $\left(\mathbf{S}_b^{(n)} - \alpha\mathbf{S}_w^{(n)}\right)$.
10:         Convergence condition: $F^{(t)}(\mathbf{U}_n) = \sum\limits_{n=1}^{N} tr\left[(\mathbf{U}_n^{(t)})^{\mathrm{T}}[\mathbf{S}_b^{(n)} - \alpha\mathbf{S}_w^{(n)}]\mathbf{U}_n^{(t)}\right],$
11: **if** $\|F^{(t)}(\mathbf{U}_n) - F^{(t-1)}(\mathbf{U}_n)\| \leq \varepsilon$, return.
12:     **end for**
13:     $\mathbf{U}_n^{(t-1)} = \mathbf{U}_n^{(t)}$, $1 \leq n \leq$ N.
14: **end for**

---

3 Experiments

In this section, we report three experiments, including depth action recognition, silhouette action recognition, and RGB object recognition. We test the performance of discriminant correlation analysis in action sequence and RGB pattern, and show the effectiveness of the proposed method. Besides, we show some important properties of our method, such as convergence, training time, and similarity matrices of classes.

3.1 Datasets and Compared Methods

3.1.1 MSR 3D Action Dataset

This dataset[1] contains 20 categories of depth actions, which are *arm waving, horizontal waving, hammer, hand catching, punching, throwing, drawing x, drawing circle, clapping, two hands waving, sideboxing, bending, forward kicking, side kicking, jogging, tennis swing, golf swing*, and *picking up and throwing*.

There are total 567 samples from 10 subjects, and each sample is performed 2~3 trials. Each action is composed of a series of frames. In order to align the image

---

[1]http://research.microsoft.com/en-us/um/people/zliu/ActionRecoRsrc/.

Fig. 4.3 Actions of MSR 3D action dataset

Fig. 4.4 ETH dataset. The objects from *left* to *right* are apple, car, cow, cup, dog, horse, pear, and tomato, respectively

sequences with one another, we first cropped and resized each frame to be 80×80, then we subsampled each action sequences to be in the size of $80 \times 80 \times 10$. The key frames are shown in Fig. 4.3.

3.1.2 KTH Action Dataset

KTH dataset [21] is employed in this experiment, and it contains action classes: running, boxing, waving, jogging, walking, and clapping. There are 90 videos with 9 people, and each person performs 10 class actions. We took the outdoor scenario in this experiment. From the videos 1,310 samples are distilled, and each sample is in the size of $90 \times 50 \times 20$ pixels. tenfold cross-validation is used for the dataset. Each time 101 samples are used for training, and 30 for testing.

3.1.3 ETH Dataset

In this experiment, we use an RGB object dataset to testify the proposed method by generalizing its application in addition to action sequence. In this experiment we employed the ETH dataset [16], which is an ideal simple dataset, to test the performance of our method. There are 3,280 RGB images belonging to 8 classes, each of which has 10 subjects. Figure 4.4 shows the samples of each class. Leave-one-out cross validation was used, and each time one subject was selected for testing, and the rest for training.

3.1.4 Methods for Comparison

We used DATER [26], MDCC [11], DNTF [30], V-TensorFace [23] (renamed as V-Tensor for simplicity) for comparison. All of them except V-tensor are discriminant tensor methods. V-Tensor is a kind of mode-driven method, which finds the *mode*-3 neighbors of tensor, while *mode*-1,2 do not. DATER applied Fisher criteria on each mode of a tensor sample directly. MDCC performs discriminant

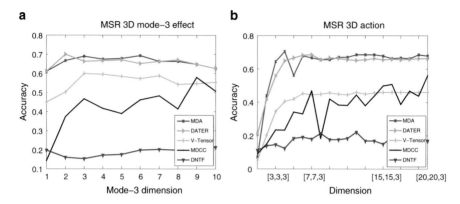

Fig. 4.5 Recognition of different methods versus different dimensions on MSR 3D action dataset. (**a**) Mode-3 dimension is changed. (**b**) Mode-3 dimension is not changed

CCA in the tensor subspace. DNTF employs discriminant non-negative matrix factorization in tensor subspace, and the key is to iteratively calculated the non-negative transformation matrices.

3.2 Results

3.2.1 MSR 3D Action Dataset

Here, the effect of dimensional reduction on the time series is tested. Figure 4.5a indicates the accuracy under various *mode*-3 dimension while the dimension of *mode*-1,2 are fixed to be 10. We can see the proposed MDA is comparable with DATER, which reflects the less influence of *mode*-3 correlation of frame series. Besides, the V-Tensor performs worse than both MDA and DATER, which also indicates that *mode*-3 manifold does not work better than Fisher criterion. Here, MDCC performs worse than MDA and V-Tensor, i.e., the similarity of *mode*-1,2 plays a negative role because of different levels of distortion by the previous cropping. Figure 4.5b shows the recognition accuracy of different methods under various dimensions. As shown in Fig. 4.5b shows, we set the *mode*-3 dimension to be 3, while the dimensions of *mode*-1,2 is increased from 1 to 20. We can see that the results of our method are comparable to the accuracy of DATER in average, which means the correlation analysis is feasible on the time series. All the results do not improve much when the dimensions are small, which indicate the optimal dimensions of subspace.

3.2.2 KTH Action Dataset

Figure 4.6a shows the effect of dimensional reduction on *mode*-3 data. We can see the accuracy is increased along with the increase of dimensions. So, next step

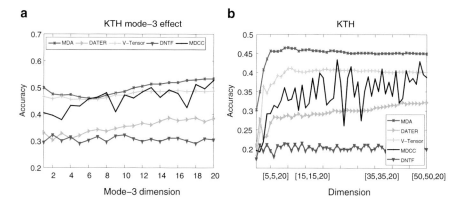

Fig. 4.6 Recognition rates w.r.t. dimension of different methods. (**a**) Mode-3 dimension is changed. (**b**) Mode-3 dimension is not changed

we will test the effect of the *mode*-1,2 dimension by fixing the *mode*-3. In Fig. 4.6b, the dimension of *mode*-3 is fixed to be 20, and we can see that MDA gets the best result within dimensions of [15,15,20], which indicates the best subspace dimension of *mode*-1,2. MDA performs better than DATER, which means the *mode*-3 correlation plays an important role in the action silhouette sequence. V-Tensor also performs better than DATER in this dataset, which indicates the effectiveness of the action sequence by *mode*-3 manifold learning. MDCC gets better results than DATER most of the time, which is another evidence for the effectiveness of mode-3 correlation. DNTF is suitable for dealing with images with rich information, like face, etc. While for silhouette with 0 and 1 value, the preserved energy is too little to perform well.

3.2.3   ETH Dataset

The recognition results are shown in Fig. 4.7, in which the projected dimensions are selected from [1,1,1] to [32,32,3]. MDA performs well even when the image dimension is small, which demonstrates that the *mode*-1,2 rank is small, and the discriminant correlations of *mode*-3 is effective. DNTF performs well in the common ETH object dataset, while poorly in the KTH dataset. This is reasonable because the pixel-based input data, the unoccluded and undamaged RGB object contains much more information than an action silhouette, which contains 0 and 1 value only.

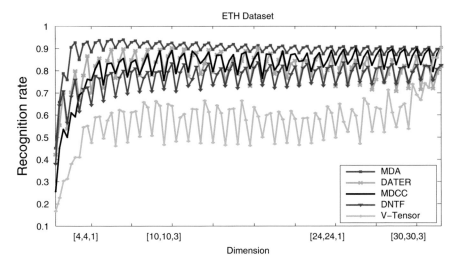

Fig. 4.7 The recognition rate of different methods versus different dimensions

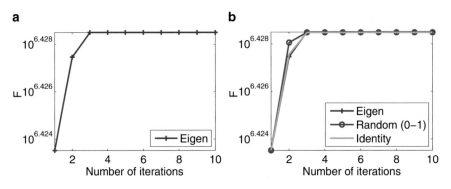

Fig. 4.8 Convergence characteristics of the optimization in MDA. (**a**) is the F value of total modes varies with the iterations. (**b**) shows the convergence to a unique maximum value with different initial values of **U**. This sub-figure indicates that MDA is convex

3.3 Properties Analysis

3.3.1 Convergence and Training Time

We analyze the convergence property and training time on KTH action dataset. The convergence character of MDA is demonstrated in Fig. 4.8 from which we can see that each experiment for learning uses a different training data set and starts with the initial value of U_n, which is composed of eigenvectors. The value of the discriminant function F (calculated as the sum of F_n) becomes stable after the first few iterations. This fast and stable convergence property is very suitable for keeping the learning

Table 4.1 Training time (second) on the KTH dataset

Method	DATER [26]	MDCC [11]	DNTF [30]	V-tensor [23]	MDA [Ours]
Time (s)	$\sim$10	$\sim$10	$\sim$5	$\sim$5	$\sim$5

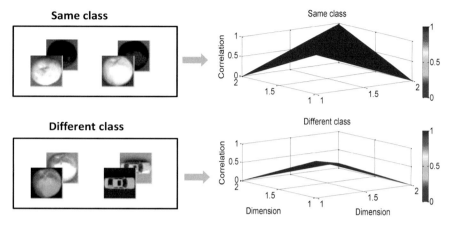

Fig. 4.9 Similarity matrices for MDA on ETH dataset. The *upper left* upper part indicates projected tensors from the same class, whose correlations illustrated in the similarity matrix shown in the *upper right*. The *lower part* shows the situation from different class. It can be intuitively seen the correlation on the second mode is smaller than that of the upper

cost low. Furthermore, as shown in Fig. 4.8b, MDA converged to the same point irrespective of different initial values of $\mathbf{U}_n$, which means that MDA is convex.

The training time of all the methods is shown in Table 4.1. The time complexity of MDA is also competitive with others.

3.3.2 Similarity Analysis

We analyze the similarity property on ETH action dataset. Figure 4.9 shows the similarity matrices of samples from the same and different classes. The *mode*-3 reduced dimension is 2. The corresponding correlations in the similarity matrices are ρ_1 and ρ_2, and $diag(\rho_1) = [1, 0.9812]$, $diag(\rho_2) = [1, 0.1372]$. Figure 4.9 illustrates that the correlation value of the same class is larger than that of different classes, which indicates that the *mode*-3 discriminant performance of MDA is reliable.

4 Summary

In this chapter we proposed a novel mode alignment method with different discriminant criteria in tensor subspace, and we name this method MDA. MDA is used to perform the dimension reduction tasks on different tensorial modes; MDA employs Fisher criteria on the first $(N - 1)$ feature modes of all the tensors to extract image features, then it updates the whole tensor sequence by considering the correlations of $(k - 1)$th, kth, and $(k + 1)$th subsets, and finally it performs discriminant analysis on the Nth mode to calculate *mode*-3 projection matrix. The proposed MDA outperforms other tensor-based methods in two commonly used action datasets: MSR action 3D dataset, KTH action dataset, and one object ETH dataset. The time complexity of MDA is lower than or equal to others, therefore it is suitable for large-scale computing. MDA can deal well with damaged action silhouettes and RGB object images in various view angles, which demonstrates its robustness. Moreover, the alternating projection procedure of MDA converges, as proved theoretically and confirmed by experiments, and finally, MDA is convex with different initial values of the transformation matrices.

References

1. Ballani, J., Grasedyck, L.: A projection method to solve linear systems in tensor format. Numer. Linear Algebra Appl. **20**(1), 27–43 (2013)
2. Belhumeur, P., Hespanha, J., Kriegman, D.: Eigenfaces vs. Fisherfaces: recognition using class specific linear projection. IEEE Trans. Pattern Anal. Mach. Intell. **19**(7), 711–720 (1997)
3. Biswas, S., Aggarwal, G., Flynn, P.J., Bowyer, K.W.: Pose-robust recognition of low-resolution face images. IEEE Trans. Pattern Anal. Mach. Intell. **35**(12), 3037–3049 (2013)
4. Fukunaga, K.: Introduction to statistical pattern recognition. Pattern Recogn. **22**(7), 833–834 (1990)
5. Gong, D., Medioni, G.: Dynamic manifold warping for view invariant action recognition. In: International Conference on Computer Vision, pp. 571–578. IEEE (2011)
6. Gong, W., Sapienza, M., Cuzzolin, F.: Fisher tensor decomposition for unconstrained gait recognition. Training **2**, 3 (2013)
7. Guo, K., Ishwar, P., Konrad, J.: Action recognition from video using feature covariance matrices. IEEE Trans. Image Process. **22**(6), 2479–2494 (2013)
8. Ho, H.T., Gopalan, R.: Model-driven domain adaptation on product manifolds for unconstrained face recognition. Int. J. Comput. Vis. **109**(1–2), 110–125 (2014)
9. Hu, H.: Enhanced gabor feature based classification using a regularized locally tensor discriminant model for multiview gait recognition. IEEE Trans. Circuits Syst. Video Technol. **23**(7), 1274–1286 (2013)
10. Huang, C.H., Yeh, Y.R., Wang, Y.C.F.: Recognizing actions across cameras by exploring the correlated subspace. In: Computer Vision–ECCV 2012. Workshops and Demonstrations, pp. 342–351. Springer (2012)
11. Jia, C.-C., et al.: Incremental multi-linear discriminant analysis using canonical correlations for action recognition. Neurocomputing **83**, 56–63 (2012)
12. Jia, C., Pang, W., Fu, Y.: Mode-driven volume analysis based on correlation of time series. In: Agapito, L., Bronstein, M.M., Rother, C. (eds.) Computer Vision - ECCV 2014 Workshops. Lecture Notes in Computer Science, vol. 8925, pp. 818–833. Springer (2015). doi:10.1007/978-3-319-16178-5_57. http://dx.doi.org/10.1007/978-3-319-16178-5_57
13. Kim, T.-K., Cipolla, R.: Canonical correlation analysis of video volume tensors for action categorization and detection. IEEE Trans. Pattern Anal. Mach. Intell. **31**(8), 1415–1428 (2009)
14. Kolda, T., Bader, B.: Tensor decompositions and applications. SIAM Rev. **51**(3), 455–500 (2009)

15. de Laat, K.F., van Norden, A.G., Gons, R.A., van Oudheusden, L.J., van Uden, I.W., Norris, D.G., Zwiers, M.P., de Leeuw, F.E.: Diffusion tensor imaging and gait in elderly persons with cerebral small vessel disease. Stroke **42**(2), 373–379 (2011)
16. Leibe, B., Schiele, B.: Analyzing appearance and contour based methods for object categorization. In: Conference on Computer Vision and Pattern Recognition, vol. 2, pp. II–409 (2003)
17. Lui, Y.M., Beveridge, J.R.: Tangent bundle for human action recognition. In: FG, pp. 97–102. IEEE (2011)
18. Miyamoto, K., Adachi, Y., Osada, T., Watanabe, T., Kimura, H.M., Setsuie, R., Miyashita, Y.: Dissociable memory traces within the macaque medial temporal lobe predict subsequent recognition performance. J. Neurosci. **34**(5), 1988–1997 (2014)
19. Nagendar, G., Bandiatmakuri, S.G., Tandarpally, M.G., Jawahar, C.: Action recognition using canonical correlation kernels. In: Asian Conference on Computer Vision, pp. 479–492 (2013)
20. Perez, E.A., Mota, V.F., Maciel, L.M., Sad, D., Vieira, M.B.: Combining gradient histograms using orientation tensors for human action recognition. In: International Conference on Pattern Recognition, pp. 3460–3463. IEEE (2012)
21. Schuldt, C., Laptev, I., Caputo, B.: Recognizing human actions: a local SVM approach. In: International Conference on Pattern Recognition, vol. 3, pp. 32–36 (2004)
22. Tao, D., Li, X., Wu, X., Maybank, S.: General tensor discriminant analysis and gabor features for gait recognition. IEEE Trans. Pattern Anal. Mach. Intell. **29**, 1700–1715 (2007)
23. Tian, C., Fan, G., Gao, X., Tian, Q.: Multiview face recognition: from tensorface to v-tensorface and k-tensorface. IEEE Trans. Syst. Man Cybern. B **42**(2), 320–333 (2012)
24. Wu, X., Wang, H., Liu, C., Jia, Y.: Cross-view action recognition over heterogeneous feature spaces. In: International Conference on Computer Vision, pp. 609–616 (2013)
25. Xue, G., Mei, L., Chen, C., Lu, Z.L., Poldrack, R., Dong, Q.: Spaced learning enhances subsequent recognition memory by reducing neural repetition suppression. J. Cogn. Neurosci. **23**(7), 1624–1633 (2011)
26. Yan, S., Xu, D., Yang, Q., Zhang, L., Tang, X., Zhang, H.: Discriminant analysis with tensor representation. In: Conference on Computer Vision and Pattern Recognition, vol. 1, pp. 526–532 (2005)
27. Yang, F., Bourdev, L., Shechtman, E., Wang, J., Metaxas, D.: Facial expression editing in video using a temporally-smooth factorization. In: Conference on Computer Vision and Pattern Recognition, pp. 861–868. IEEE (2012)
28. Youn, J., Cho, J.W., Lee, W.Y., Kim, G.M., Kim, S.T., Kim, H.T.: Diffusion tensor imaging of freezing of gait in patients with white matter changes. Mov. Disord. **27**(6), 760–764 (2012)
29. Yu, Z.Z., Jia, C.C., Pang, W., Zhang, C.Y., Zhong, L.H.: Tensor discriminant analysis with multiscale features for action modeling and categorization. IEEE Signal Process Lett. **19**(2), 95–98 (2012)
30. Zafeiriou, S.: Discriminant nonnegative tensor factorization algorithms. IEEE Trans. Neural Netw. **20**(2), 217–235 (2009)

Chapter 5
RGB-D Action Recognition

Chengcheng Jia, Yu Kong, Zhengming Ding, and Yun Fu

1 Introduction

1.1 RGB-D Action Recognition

For human action recognition task, the traditional methods are based on the RGB data captured by webcam, while the popular RGB action databases[1] are like UT-Interaction [34], UCF Sports [32], UCF 101 [38], KTH [35], and Hollywood database [16]. Meanwhile, methods [1, 9, 17, 31, 48] that were designed for these RGB action databases cannot utilize the rich 3D-structural information to reduce large intra-class variations.

Conventional action recognition methods focused on RGB videos which can roughly be divided into two categories: low-level feature-based methods and

---

[1]In this work, RGB action database means the one that is captured by a conventional RGB camera. Only RGB data are available in the RGB action database; depth data are not available. RGB-D action database is the one that is captured by a RGB-D camera. Both RGB data and depth data are available in the RGB-D action database.

C. Jia (✉) • Y. Kong • Z. Ding
Department of Electrical and Computer Engineering, Northeastern University, 360 Huntington Avenue, Boston, MA 02115, USA
e-mail: jia.ch@husky.neu.edu; yukong@ece.neu.edu; allanding@coe.neu.edu

Y. Fu
Department of Electrical and Computer Engineering and College of Computer and Information Science (Affiliated), Northeastern University, 360 Huntington Avenue, Boston, MA 02115, USA
e-mail: yunfu@ece.neu.edu

© Springer International Publishing Switzerland 2016
Y. Fu (ed.), *Human Activity Recognition and Prediction*,
DOI 10.1007/978-3-319-27004-3_5

mid-level knowledge-based methods. Low-level feature-based methods [4, 11, 15, 18, 35] use spatiotemporal interest points, body shape feature [23], structure information [33], key poses [31], etc., to represent human actions. These methods rely on hand-crafted features, where the actions directly learn from. Recent work also shows that action features can be learned using deep learning techniques [9]. Mid-level features, such as attributes [25], optical flow [45], semantic descriptions [13], context [2], and relations [40], are learned from low-level features and then used for the recognition task. The learned mid-level features can be considered as knowledge discovered from the same database used for training or having been specified by experts.

Even with the state-of-the-art RGB action recognition methods having achieved high performance on some action classes, they often fail to understand actions in more challenging scenarios (e.g., occluded body parts) due to the lack of 3D structural information. On the other hand, several methods [6, 20, 47, 49] have been proposed to utilize 3D-structural information, but require real RGB-D databases, and cannot be directly applied to RGB databases.

Due to the recent advent of the cost-effective Kinect sensors, action recognition from RGB-D cameras has received an increasing interest throughout the computer vision community. Compared with conventional RGB cameras, Kinect sensors provide depth information, computing 3D-structural information of the entire scene. The 3D-structural information can facilitate the recognition task by simplifying intra-class motion variation and removing cluttered background noise. Recent work [6, 46, 47, 49] has shown that the performance of recognition systems can be improved by applying depth information to yield RGB-D databases (e.g., MSR DailyAction3D database [48] and MSR PairAction3D database [27]).

For the human action representation, there are many previous studies that used tensor for high-dimensional data representation [39], in order to distinguish their various factors (e.g., identity, expression, illumination, color in face recognition [29, 41], object classification [42], and action video recognition [10, 19, 43]). Motivated by these methods, a tensor could be employed to represent an action sequence. Specifically, a RGB or depth action video is represented as a third-order tensor, while a RGB-D action video is composed of two third-order tensors. It can also be constructed as a fourth-order tensor: when there is a "missing" modality (e.g., RGB or depth), the fourth-order tensor can be truncated to a third-order tensor.

1.2 Our Contribution

In this chapter, we propose a novel action recognition method for learning the "missing" depth information[2] that is not included in conventional RGB action

---

[2]In this work, the missing depth data means that they are not included in the conventional RGB action databases, or we intentionally remove them from a RGB-D database to build a RGB database for evaluation.

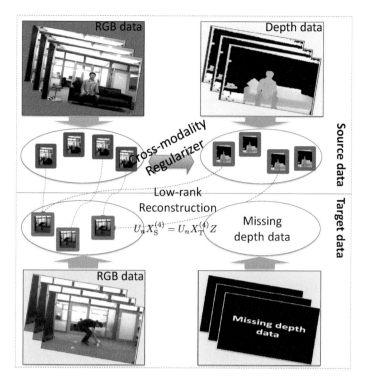

Fig. 5.1 Our method is able to transfer knowledge from the RGB-D dataset to the RGB dataset, so missing depth information can be compensated for. Using a latent low-rank constraint $U_nX_S = U_nX_TZ + LU_nX_S + E$, the missing depth information in the target is uncoverable. A novel cross-modality regularizer is added to couple the RGB and depth modalities to properly align and transfer more depth information to that of the target

databases (Fig. 5.1) to further improve the recognition performance. In the first step, we utilize a RGB-D database as the source data and learn the knowledge about the correlations between the two modalities: RGB data and depth data of the source data. Second, this knowledge is then transferred to the target data, which only has RGB information. Consequently, depth information can be compensated for in the target data. Moreover, a cross-modality regularizer is added to the two modalities of source data so that RGB and depth information can be well coupled and more depth information can be transferred to the target database. Finally, on the target database, our method can extract useful 3D-structural information, improving the performance over methods that only uses the real RGB data.

There are several properties of our method worthy emphasizing here. First, we incorporate the knowledge about the correlations between RGB and depth data into our model. The use of this rich knowledge allows us to use additional source of information for recognition. Even though this knowledge does not exist in the target training and testing data, we can still learn this knowledge from an additional RGB-D source data and then transfer it to the RGB target data. Second, we bridge the gap between RGB database and RGB-D database. Previous methods treat them as different application subjects, i.e., action recognition [1, 17, 48] and RGB-D action recognition [6, 20, 47, 49]. By comparison, in this work, we show that RGB-D database can be used as strong depth information for the RGB database. Consequently, conventional RGB action databases can be reused as "RGB-D" databases and provide action recognition methods with rich depth information fused within. Third, compared with attribute-based methods [13, 14, 25] for recovering missing training samples, the knowledge in our work is data driven and is automatically discovered from data rather than being manually specified by experts. This helps us mine useful knowledge and avoid subjective labeling in an automated fashion.

2 Related Work

2.1 Action Recognition

The traditional action recognition task is based on RGB data, which has homogeneous knowledge in the same database, and achieved high performance on some classes with low-level and mid-level features. However, they often fail to understand actions in more challenging scenarios.

Recently, due to the advent of the cost-effective Kinect sensor, researchers have put lots of attentions to the recognizing actions from RGB-D data [6, 22, 27, 46, 47, 49]. Compared with conventional RGB data, the additional depth information has several advantages to handle complex scenarios. First, it allows us to capture 3D structural information, which is very useful in removing background information and simplifying intra-class motion variations. Second, it eliminates the effect of illumination and color variation.

In our work, depth information is available in the target training and testing databases. First, we use additional RGB-D data as the source database and then learn the correlations between RGB data and depth data. Third, the learned correlation knowledge is transferred to the target database where the depth information does not exist. With the learned depth information, the performance on the target RGB data improves from when only the RGB data in the target data are used.

2.2 Transfer Learning

In this chapter, we mainly focus on using RGB-D data to help to improve the performance of recognizing RGB data; therefore, we introduce the widely used transfer learning theory to help to understand our method.

Transfer learning techniques attract lots of attention those recent years, which aim to transfer existing, well-instituted knowledge to the new datasets or new problems. Generally, transfer learning can be categorized by domains and tasks. There are two categories of transfer learning: first, inductive transfer learning is defined by the same data domain, but by different tasks. Second, transductive transfer learning [8, 36] uses different data domains for the same task. In transfer learning problems, target data and source data should share common properties, as the information in the source domain is available for the target domain to reference. Additional detailed references are available in the survey of transfer learning [28].

More recently, a low-rank constraint was introduced to transfer learning [8, 36]. Low-rank representation (LRR) [24, 26] can discover the structure information of the data, especially the data coming from various subspaces. LRR has several purposes: first, it finds the lowest-rank representation among all the data, which can be represented as the linear combinations of the basis. Such a LRR can uncover the dataset in its entire global structure, grouping the correlated candidates into one cluster. Second, LRR handles more noise and gross corruption of data than other traditional methods that only consider the Gaussian noise. Third, Liu et al. proposed a latent low-rank representation (LatLRR) method to handle the limited observed data problem [21]. LatLRR discovers the latent information of the unobserved data from the observed, making the recovery more stable.

Even more recently, a low-rank constraint was used in transfer learning. LTSL [36] and LRDAP [8] are two typical transfer learning methods that include the low-rank constraint. LTSL aims to find a common subspace where the source data can well represent the target in the low-rank framework. LRDAP aims to find a rotation on source domain data to be represented by the target domain in the low-rank framework. LRDAP considers that the rotated source data can be used to test the target data when the rotated source data can be reconstructed in the target domain.

Different from the previous work, our method solve missing modality problem by transferring depth, from RGB-D source data to the RGB target data in tensor framework. By introducing a new cross-modality regularizer on the source data, our method can couple the two RGB and depth modalities to transfer more depth information into the target domain to compensate for the missing depth information in the target data.

3   Proposed Method

In this section, we introduce a new RGB-D action transfer learning method for recognition. First, we introduce the framework and then give the details of the proposed method with three major parts: *latent problem formulation, cross-modality regularizer, and solving objective function.* Third, complexity is analyzed in the following, and more properties will be discussed in this section.

3.1   Framework Overview

We overview the framework in this section. First, we introduce the data representation, and then we explain the workflow. Suppose both RGB and depth data is a
third-order tensor, and the whole dataset is taken as a *fourth-order* tensor. In the
source domain, there is RGB-D data used as auxiliary data. While in target domain,
there is only one modality (RGB/depth) data, whose performance is effected by the
auxiliary data.

   We apply latent low-rank transfer learning on the *fourth mode* of the tensor data.
First, a cross-modality regularizer on the two modalities of source database is used
to properly align the two modalities. In this way, the correlation of RGB and depth
channels could be transferred to target domain, to help to discover depth information
in the target/test. Second, the latent low-rank constraint is used to recover the depth
information in the target domain from the source domain with complete modalities,
to help compensate for the missing modality in the target database.

3.2   Preliminary

We introduce the fundamental theory of tensor decomposition in this section. A
multidimensional array $\mathcal{X} \in \mathbb{R}^{I_1 \times \dots \times I_n \times \dots \times I_N}$ is called an *Nth-order* tensor, where I_n
is the size of n-th dimension. For example, a vector $x \in \mathbb{R}^{I_1}$ can be taken as a *first-
order* tensor, and a matrix $X \in \mathbb{R}^{I_1 \times I_2}$ is called a *second-order* tensor. Each element
of $\mathcal{X}$ is represented as $\mathcal{X}_{i_1 \dots i_n \dots i_N}$, where i_n is the index of n-th dimension (or mode).
The n-th mode of $\mathcal{X}$ is of size I_n, while each fiber $x \in \mathbb{R}^{I_n}$ of n-th mode is a *mode-n*
vector [12].

Definition 1 (Mode-n Unfolding). An Nth-order tensor $\mathcal{X}$ can be stretched by
fixing the mode-n vectors to be a matrix $X_{(n)} \in \mathbb{R}^{I_n \times (I_1 \cdot I_2 \dots I_{n-1} \cdot I_{n+1} \dots I_N)}$, where $\cdot$ is
used for scalar product.

Definition 2 (Tucker Decomposition). Given a tensor $\mathcal{X} \in \mathbb{R}^{I_1 \times \dots \times I_N}$, and projection matrices $U_n \in \mathbb{R}^{I_n \times J_n} (1 \leq n \leq N)$, the decomposition is performed by

$$\mathcal{S} = \mathcal{X} \times_1 U_1 \times_2 U_2 \dots \times_n U_n \dots \times_N U_N, \qquad (5.1)$$

where $\mathcal{X} = \mathcal{S} \times_1 U_1^T \times_2 U_2^T \dots \times_n U_n^T \dots \times_N U_N^T$, $\times_n$ indicates mode-n product.
$\mathcal{S} \in \mathbb{R}^{J_1 \times J_2 \times \dots \times J_N}$ is called the core tensor.

   The illustration of the *mode-n* unfolding of $\mathcal{X}$ is shown in Fig. 5.2. By unfolding
along *x-axis*, we get a matrix ordered by images side by side, and along *y-axis*, all
the images are ordered in a column, while along *z-axis*, we put the cross section side
by side.

3.3 Latent Problem Formulation

Given two action databases in fourth-order tensor representation, $\mathcal{X}_S \in \mathbb{R}^{(I_1 \times I_2 \times I_3) \times I_S}$ and $\mathcal{X}_T \in \mathbb{R}^{(I_1 \times I_2 \times I_3) \times I_T}$, where I_1 and I_2 mean the row and column dimensions of one frame, I_3 indicates the number of frames in one action sample, and I_S, I_T are the numbers of action samples in source and target databases, respectively. Each database has two modalities, RGB and depth, which are combined in the fourth order of the data tensor, i.e., $\mathcal{X}_S = [\mathcal{X}_{S\text{-RGB}}, \mathcal{X}_{S\text{-D}}]$ and $\mathcal{X}_T = [\mathcal{X}_{T\text{-RGB}}, \mathcal{X}_{T\text{-D}}]$. Traditional transfer learning methods would consider the transfer between modalities within one database (e.g., $\mathcal{X}_{S\text{-RGB}} \rightarrow \mathcal{X}_{S\text{-D}}$ and $\mathcal{X}_{T\text{-RGB}} \rightarrow \mathcal{X}_{T\text{-D}}$) or the transfer between two databases in one modality (e.g., $\mathcal{X}_{S\text{-RGB}} \rightarrow \mathcal{X}_{T\text{-RGB}}$ and $\mathcal{X}_{S\text{-D}} \rightarrow \mathcal{X}_{T\text{-D}}$). However, when one modality of the target is missing, both the kinds of traditional transfer learning would fail. How can we uncover the information of the lost modality? In this chapter, we propose a latent low-rank tensor transfer learning by introducing a graph regularizer to couple RGB and depth information in the source database and then transfer to the target database. We consider the transfer $\mathcal{X}_S \rightarrow \mathcal{X}_T$, where the source has two modalities, while the target contains only one modality.

Since $\mathcal{X}_S$ and $\mathcal{X}_T$ are two different databases lying in different spaces, $\mathcal{X}_S \nsubseteq \mathcal{X}_T$. Suppose the source data $\mathcal{X}_S$ and target data $\mathcal{X}_T$ in the same space, both projected with U_n ($n = 1, 2, 3$) on the first three modes of databases, i.e., $U_n \mathcal{X}_S \subseteq U_n \mathcal{X}_T$. First, we assume $\mathcal{X}_{T\text{-D}}$ is known and unfold both the tensors in the fourth order to get $U_n X_S^{(4)}, U_n X_T^{(4)}$; see Fig. 5.2. As many previous researches have shown, LRR in transfer learning helps preserve the locality in reconstruction and find the block structure of the source data. We define the low-rank tensor transfer learning with unfolded tensor data as

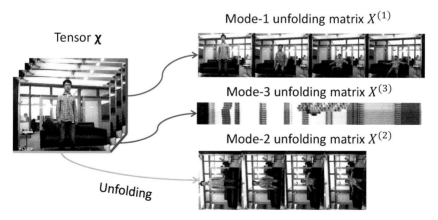

Fig. 5.2 Illustration of tensor *mode-n* unfolding. An action sequence is represented as a 3D tensor $\mathcal{X}$, and we get *mode-n* unfolding matrix $X^{(n)}$ along its *x-, y-,* and *z-axis*

$$\min_{Z} \|Z\|_*,$$
$$\text{s.t. } U_n X_S^{(4)} = U_n X_T^{(4)} Z, \tag{5.2}$$
$$U_n^T U_n = I, \quad n = 1, 2, 3,$$

where $\|\cdot\|_*$ is the nuclear norm and Z is called "low-rank representations" of source data $X_S^{(4)}$ with respect to the target data $X_T^{(4)}$. We focus on the unfolded tensor in the fourth order, so we omit the superscript ".$^{(4)}$" for simplicity, since the source and target data are well aligned after projected by U_n ($n = 1, 2, 3$).

We propose to find the relationship between X_S and X_T via SVD, i.e., $U_n[X_S, X_T] = H\Sigma V^T$, where $V = [V_S; V_T]$ by row partition. Therefore, $U_n[X_S, X_T] = H\Sigma[V_S; V_T]^T = [H\Sigma V_S^T; H\Sigma V_T^T]$, and $U_n X_S = H\Sigma V_S^T$, $U_n X_T = H\Sigma V_T^T$ can be derived. The above constraint can be written as $H\Sigma V_S^T = H\Sigma V_T^T Z$. According to Theorem 3.1 [21], the optimal LRR Z_* can be found as follows: $Z_* = V_T V_S^T = [V_{T\text{-RGB}}; V_{T\text{-D}}] V_S^T$, where V_T has been row partitioned into $V_{T\text{-RGB}}$ and $V_{T\text{-D}}$. Then the constrained part can be rewritten as

$$\begin{aligned}
U_n X_S &= U_n X_T Z_* \\
&= U_n[X_{T\text{-RGB}}, X_{T\text{-D}}] Z_* \\
&= U_n[X_{T\text{-RGB}}, X_{T\text{-D}}][V_{T\text{-RGB}}; V_{T\text{-D}}] V_S^T \\
&= U_n X_{T\text{-RGB}} V_{T\text{-RGB}} V_S^T + U_n X_{T\text{-D}} V_{T\text{-D}} V_S^T \\
&= U_n X_{T\text{-RGB}} (V_{T\text{-RGB}} V_S^T) + U\Sigma V_{T\text{-D}}^T V_{T\text{-D}} V_S^T \\
&= U_n X_{T\text{-RGB}} \tilde{Z} + (U\Sigma V_{T\text{-D}}^T V_{T\text{-D}} \Sigma^{-1} U^T) U_n X_S \\
&= U_n X_{T\text{-RGB}} \tilde{Z} + \tilde{L} U_n X_S,
\end{aligned} \tag{5.3}$$

where $\tilde{Z}$ indicates the low-rank structure of source data on the target data only with RGB and $\tilde{L} = U\Sigma V_{T\text{-D}}^T V_{T\text{-D}} \Sigma^{-1} U^T$ should also be low rank, as $V_{T\text{-D}}^T V_{T\text{-D}}$ aims to uncover the block structure of $U_n X_{T\text{-D}}$.

In reality, the data is sometimes corrupted. So we need to add an error term as done previously in [24, 26, 36]; then the general model in Eq. (5.2) is represented as

$$\min_{Z,L,E} \|Z\|_* + \|L\|_* + \lambda \|E\|_{2,1}$$
$$\text{s.t. } U_n X_S = U_n X_T Z + L U_n X_S + E, \tag{5.4}$$
$$U_n^T U_n = I, \quad n = 1, 2, 3,$$

where $\|E\|_{2,1} = \sum_{i=1}^{I_S} \sqrt{\sum_{j=1}^{(I_1 \times I_2 \times I_3)} ([E]_{ij})^2}$ is $L_{2,1}$ norm, which can make E sample specific (column sparsity), resulting in the formula detecting some outliers. $\lambda > 0$ is the parameter to balance the error.

As shown in [21], LatLRR provides a way to reconstruct the hidden data matrix from two directions: column and row of the observed data. Our algorithm also has such an advantage. The source data $U_n X_S$ is reconstructed from both the fourth order of $U_n X_T$ and the first three orders of $U_n X_S$. That is, when some data in the target domain is missing, $U_n X_S$ will make sense in reconstruction and uncover the latent information of the missing data. When solving the previous problem, the objective testing modality is the missing part of the target data. We can then uncover more latent information from the source data in transfer subspace learning.

3.4 Cross-Modality Regularizer

The latent low-rank tensor model can uncover the lost modality of the target using the source dataset. In this section, we introduce a cross-modality regularizer to capture the modality information in the source data, so that more depth information from the source dataset can be transferred to the target dataset. The learned Z can be treated as a new LRR of the projected source data $U_n X_S$, which is a combination of RGB data Z_{RGB} and depth data Z_D. Our goal is to couple the different RGB and depth modalities of the same class by the similar low-rank coefficients. We define z_i as the i-th column of Z, which correlates with the i-th sample x_i of X_S. According to this property, we introduce a novel regularization term.

$$\min_{z_i, z_j} \sum_{i=1}^{l_S} \sum_{j=1}^{l_S} (z_i - z_j)^2 w_{ij}, \tag{5.5}$$
$$\forall i, j, w_{ij} \in \mathcal{W},$$

where $w_{ij} = 1$, if x_i and x_j have the same label; $w_{ij} = 0$ otherwise. From this formulation, we saw that Eq. (5.5) enforces z_i and z_j to be similar if w_{ij} is 1. Mathematically, Eq. (5.5) can be rewritten as $\mathrm{tr}(Z^T \mathcal{L} Z)$, where $\mathcal{L} = \mathcal{D} - \mathcal{W}$ and $\mathcal{D}$ is a diagonal matrix with the rows sum of $\mathcal{W}$ as the element. It can be seen that $\mathcal{L}$ is very similar to graph Laplacian, which has been extensively used in spectral clustering [44] and graph embedding [50]. However, different from them, the proposed term $\mathcal{W}$, as it carries discriminative label information between modalities for source data, where RGB and depth data with the same label are coupled. By adding the regularization term $\mathrm{tr}(Z^T \mathcal{L} Z)$ to Eq. (5.5), our final objective function results as follows:

$$\min_{Z,L,E} \|Z\|_* + \|L\|_* + \lambda \|E\|_{2,1} + \frac{\beta}{2} \mathrm{tr}(Z^T \mathcal{L} Z),$$
$$\text{s.t. } U_n X_S = U_n X_T Z + L U_n X_S + E, \tag{5.6}$$
$$U_n^T U_n = I, \quad n = 1, 2, 3,$$

where $\beta \geq 0$ is the balanced parameter. When $\beta = 0$, there is no couple between the two modalities in the source data.

With the above formula, two modalities of the source dataset can be well coupled when reconstructed by the target data with a low-rank constraint. Therefore, the information of the lost modality in target dataset can be more uncovered using the complete modality source dataset in transfer learning. In the experiment, we show that the cross-modality regularizer can help with transferring the depth information of source data to uncover the missing depth modality of the target data.

3.5 Solving Objective Function

To solve Eq. (5.6), we rewrite the objective function by introducing some auxiliary matrices as

$$
\begin{aligned}
\min_{\substack{K,Z,W,\\ L,E}} \ & \|K\|_* + \|W\|_* + \tfrac{\beta}{2}\mathrm{tr}(Z^\mathrm{T}\mathcal{L}Z) + \lambda\|E\|_{2,1}, \\
\text{s.t.} \ & U_n X_\mathrm{S} = U_n X_\mathrm{T} Z + L U_n X_\mathrm{S} + E, \\
& Z = K, \quad L = W, \quad U_n^\mathrm{T} U_n = \mathrm{I}, \quad n = 1,2,3.
\end{aligned}
\tag{5.7}
$$

Augmented Lagrangian multiplier [7, 30] is applied to achieve better convergence of Eq. (5.7); the augmented Lagrangian function is

$$
\begin{aligned}
\min_{\substack{K,W,Z,L,\\ U_n,E}} \ & \|K\|_* + \|W\|_* + \lambda\|E\|_{2,1} \\
& + \mathrm{tr}[Y_1^\mathrm{T}(U_n X_\mathrm{S} - U_n X_\mathrm{T} Z - L U_n X_\mathrm{S} - E)] \\
& + \mathrm{tr}[Y_2^\mathrm{T}(Z - K)] + \mathrm{tr}[Y_3^\mathrm{T}(L - W)] \\
& + \tfrac{\beta}{2}\mathrm{tr}(Z^\mathrm{T}\mathcal{L}Z) + \tfrac{\mu}{2}\left[\|Z - K\|_\mathrm{F}^2 + \|L - W\|_\mathrm{F}^2\right. \\
& \left. + \|U_n X_\mathrm{S} - U_n X_\mathrm{T} Z - L U_n X_\mathrm{S} - E\|_\mathrm{F}^2\right],
\end{aligned}
\tag{5.8}
$$

where Y_1, Y_2, Y_3 are Lagrange multipliers and $\mu > 0$ is a penalty parameter. All the variables in Eq. (5.8) are difficult to calculate jointly; however, we can optimize them one by one in an iterative manner. We use the Augmented Lagrangian Multiplier [7, 30] for the above problem, which converges well even when some of the data is not smooth. The procedure can be found in [10].

3.6 Discussion

We discuss some advantages of our method in this section. By borrowing an auxiliary dataset, with both RGB and depth data, our method can transfer the complete modality information from the source domain to the target. We highlight some important properties of our method and show the differences comparing with existing methods.

Uncover Missing Depth Information in the Target Data The latent low-rank constraint assumes that the transformed RGB-D source data can be represented by a transformed RGB target data plus latent information (missing depth information). This allows us to discover the missing depth information and then use it as additional cue for recognition.

Coupling Two Modalities We propose a cross-modality regularizer on the RGB and depth of source data to couple the two modalities. Therefore, they can be well aligned and transfer more depth information to the target.

Capturing Structure Information Our method captures both RGB and depth information. In addition, our tensor-based method elegantly captures spatial geometric information.

Compared with traditional transfer learning methods, such as LTSL and GFK, they do not consider the missing information in the target, as they only consider the transferring between modalities within one dataset and transfer between two datasets in one modality. Compared with L^2TSL [3], which also considers the missing modality information, our method is tensor based; therefore, our method can uncover more time spatial information from the action video. To the best of our knowledge, we are the first to consider transfer learning to uncover the depth information to help action recognition. Our novel cross-modality regularizer to compensate for the diversity of the two modalities in source data would help transfer more depth information to the target, so that our method can improve action recognition with only RGB data.

4 Experiments

In this section, we first introduce the experimental database and setting and then compare other transfer learning methods with our method; finally, we introduce some properties of our method. We classify the test samples of various modalities via borrowing the missing information from the source database. Both the source and target databases are used for training, while the target database is used as reference for testing. In the source database, the RGB-D data is used throughout every experiment.

4.1 Databases

In the experimental section, we use two action databases to evaluate the proposed method: MSR daily action 3D database and MSR pair action 3D database.[3] Both have two modalities, RGB images and depth images.

---

[3] http://research.microsoft.com/en-us/um/people/zliu/ActionRecoRsrc/.

Fig. 5.3 RGB sequence of (**a**) MSR daily action 3D database and (**b**) MSR pair action 3D database

In the MSR daily action database, there are 16 categories of actions: *drinking, eating, reading book, calling cell phone, writing on a chapter, using laptop, using vacuum cleaner, cheering up, sitting still, tossing chapter, playing game, lying down on sofa, walking, playing guitar, standing up, and sitting down.* All these actions are performed by ten subjects; each performs every action twice. There are 320 RGB samples and 320 depth samples. Figures 5.3a and 5.4a show the RGB and depth sequence of one sitting action.

In the MSR pair action database, there are six pairs of actions: *picking up a box/putting down a box, lifting a box/placing a box, pushing a chair/pulling a chair, wearing a hat/taking off a hat, putting on a backpack/taking off a backpack,* and *sticking/removing a poster.* There are a total of 360 RGB samples and 360 depth action samples. There are ten subjects performing three trails for each action, and the first five subjects are for testing. Figure 5.3b shows the RGB sequence of pushing a chair action.

To unify the tensor size, we extracted ten frames from each video sequence at specific intervals. We use the whole frame to extract feature instead of tracking, because (1) the background of the video is static and (2) we apply HOG descriptors to extract feature. Thus, the background will not contribute to the feature too much and human tracking is not necessary here.

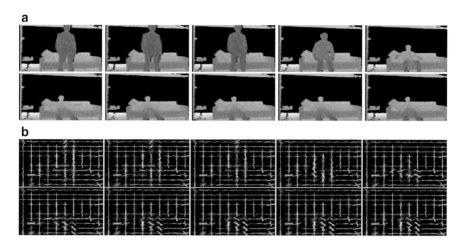

Fig. 5.4 MSR daily action 3D database, (**a**) is the depth sequence of sitting, (**b**) is the HOG feature of RGB sequence

In both databases, RGB and depth action samples were subsampled to be $80 \times 80 \times 10$, whose HOG feature[4] was exacted to represent the action samples. The HOG feature of source RGB data is shown in Fig. 5.4b.

4.2   Experimental Settings

4.2.1   MSR Pair/Daily Action Recognition

For MSR pair action recognition, **Source database:** MSR daily action 3D database, which contains 16 action categories by 10 subjects with 640 samples, composed of 320 RGB samples and 320 corresponding depth samples. **Target database:** MSR pair action 3D database, which includes 12 interaction categories by 10 subjects with 720 samples, contains 360 RGB samples and 360 depth samples. We separate the target database into two parts for training and testing, respectively. Therefore, we can test each part reversely by alternating orders, as shown in Case 1 and Case 2. For MSR daily action recognition, the source and databases are reversed, and the corresponding training setting is shown in Case 3 and Case 4.

Case 1:   The actions of the first five subjects are used for training in the target database, while the last five subjects for testing.

Case 2:   The actions by the last five subjects are employed for training, while the rest for testing.

---

[4]http://www.vlfeat.org/overview/hog.html.

Case 3:   The actions of the first five subjects are used for training in the target
database, while the last five subjects for testing.

Case 4:   The actions by the last five subjects are employed for training, while the
rest for testing.

4.3   Compared Results

In this section, we mainly use three compared methods: LTSL [37], GFK [5], and
HOG-KNN. LTSL learns a common subspace by transfer learning between two
domains for objective and face recognition, which is performed through a low-rank
constraint. GFK calculates a common subspace by finding the correlation of the
two subspaces of source and target domains, which is an unsupervised method for
objective classification. The HOG-KNN does not use transfer learning as a baseline,
i.e., the source database is not used in the training phase.

In this section, we first introduce five different settings and then give some
properties of our method, like convergence and training time.

4.3.1   Tests of Training Settings

In this section, LTSL [37], GFK [5], and HOG-KNN are used for comparison in
different test settings. We adopt five testing groups in the target database, while the
source domain and target domain are employed for training (the left three parts), and
the new samples in target database (the fourth part) are used for testing, as Table 5.1
shows.

Test 1   In target domain, we set Training–Testing data: **RGB–RGB** tensor samples,
to testify the performance of the depth information transferred from the source
database X_S on the RGB sample classification.

Left of Table 5.2 shows the accuracy results of all the methods. It can be seen
that our result is around 10 % higher than the original space, which indicates that
it works when transferring depth information for RGB action testing. While the
LTSL and GFK are lower than the HOG-KNN, meaning the two methods have

Table 5.1 Training and
testing settings for source and
target domain

Tests	Source domain		Target domain	
Test 1	*RGB*	*Depth*	*RGB*	RGB
Test 2	*RGB*	*Depth*	*Depth*	Depth
Test 3	*RGB*	*Depth*	*RGB*	Depth
Test 4	*RGB*	*Depth*	*Depth*	RGB
Test 5	*RGB*	*Depth*	*RGB*	RGB-D
	Training			Testing

Table 5.2 Accuracy (%) of left: Test 1 and right: Test 2

Test 1: RGB–RGB					Test 2: Depth–Depth				
Methods	Case 1	Case 2	Case 3	Case 4	Methods	Case 1	Case 2	Case 3	Case 4
GFK [5]	64.46	61.67	20.63	20.63	GFK [5]	71.59	74.58	36.25	39.38
LTSL [37]	57.85	46.67	6.88	5.63	LTSL [37]	60.80	69.49	5.63	5.63
HOG-KNN	76.03	64.17	19.38	18.13	HOG-KNN	90.91	89.27	34.38	38.75
Ours	**86.78**	**90.00**	**40.00**	**33.75**	Ours	**91.48**	**92.09**	**40.00**	**41.25**

Table 5.3 Accuracy (%) of left: Test 3 and right: Test 4

Test 3: RGB-Depth					Test 4: Depth-RGB				
Methods	Case 1	Case 2	Case 3	Case 4	Methods	Case 1	Case 2	Case 3	Case 4
GFK [5]	18.18	11.86	22.50	16.25	GFK [5]	7.44	19.17	15.63	14.38
LTSL [37]	10.80	7.34	6.88	5.00	LTSL [37]	7.44	12.50	5.63	5.00
HOG-KNN	22.73	24.29	28.75	26.25	HOG-KNN	12.40	22.50	17.50	16.88
Ours	**35.23**	**31.07**	**29.38**	**34.38**	Ours	**23.14**	**23.33**	**35.00**	**31.88**

transferred negative knowledge about depth and are not suitable for RGB-D action recognition. The result is reasonable because (1) in the object/face recognition, there is NO complex background in the data, so the environmental interference is very low compared to the human action data; (2) in the action recognition, there is much larger diversity between two categories, which is different from the object/face recognition containing the similar knowledge between two subjects.

Test 2 We set Training–Testing data: **Depth–Depth** tensor samples. Conversely, we also want to know if RGB knowledge is transferred from the source database; would it be helpful to improve the accuracy of the depth image recognition? So we use the depth samples as the target data for training and other depth samples for testing.

Right of Table 5.2 shows the results of depth testing, from which we can see the depth samples perform better than the RGB samples for recognition in Test 1. The accuracies of our method are shown in bold values, which are higher than others, And compared with the Test 1, the transferred RGB information had minimal impact for improving the performance when compared with HOG-KNN.

Test 3 We set Training–Testing data: **RGB–Depth** tensor samples. As we know, the RGB and depth images are two different modalities, and the matching degree will be low between them. However, if we transfer the knowledge of one modality to another, it makes sense that the matching degree between different modalities will increase, as the sharing information is transferred from the other database. Left of Table 5.3 shows the results of all the compared methods. We can see the accuracy is not as high as Test 1 and 2, because of the matching between the different modalities. However, it shows that the transferred depth knowledge helps to improve the performance by more than 10 % compared with the original space.

Test 4 We set Training–Testing data: **Depth–RGB** tensor samples. Considering Test 3, we transferred some RGB information from source database to compensate for the existing depth samples and to see the performance on the RGB action testing. Right of Table 5.3 shows the results of RGB classification by the depth images. The accuracies of our method are shown in bold values, which are higher than others. We can see the transferred RGB information helped improve by more than 10 %, but it shows that compensating depth knowledge in Test 3 works better than transferring RGB information, as done for this test.

Test 5 We set Training–Testing data: **RGB–RGB & Depth (RGB-D)** tensor samples. With RGB samples in the target database, can it be employed for new RGB-D samples classification? We suppose to generate the new samples containing both RGB and depth knowledge in the subspace, which align them with the testing RGB-D sample. All the compared results for the RGB-D action recognition are shown in Table 5.4. The accuracies of our method are shown in bold values, which are higher than others. It shows that we can make use of the existing RGB images to classify the new RGB-D samples and achieve better performance by transferring the depth knowledge from other known RGB-D databases.

According to the five tests presented, all the tables show that our method performs better than the original space, which means the transfer learning between the two modalities does work for action recognition. And some other transfer methods, which work well on the RGB objective and face recognition, may transfer the negative knowledge of the RGB or depth action samples (Fig. 5.5).

Table 5.4 Accuracy (%) of Test 5

Test 5: RGB-(RGB-D)

Methods	Case 1	Case 2	Case 3	Case 4
GFK [5]	37.04	31.99	21.56	18.44
LTSL [37]	23.23	23.23	6.88	5.31
HOG-KNN	42.09	40.40	24.06	22.19
Ours	**51.85**	**51.52**	**34.38**	**33.44**

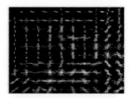

Fig. 5.5 Illustration of mode-3 feature obtained by U_3. *Left*: RGB samples; *Medium*: Corresponding HOG feature; *Right*: mode-3 feature

Fig. 5.6 Mode-n
($n = 1, 2, 3$) error with
different number of iterations

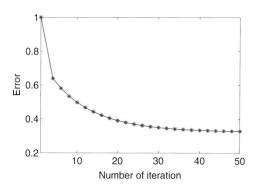

Table 5.5 Training time cost
of methods

Methods	GFK [5]	LTSL [37]	Ours
Training time	5 min	10 min	0.5 min

4.3.2 Convergence and Training Time

We use the [Test 1: Case 1] introduced in Sect. 4.2.1 to analyze the properties of the convergence and constraint term. Figure 5.6 shows the error for a different number of iterations. The convergence condition is given in Algorithm 1. We can see the curves are stable within 50 iterations.

We also compare our method with GFK [5] and LTSL [37] on the training time. We use MSR daily action 3D as source data and MSR pair action 3D as target data. All the tests are run on a PC (Intel Xeon E5 2650 CPU at 2.00 GHz and 128 GB memory). Results shown in Table 5.5 indicate that our method is significantly faster than the GFK method and the LTSL method. Our method spent 0.5 min training, which is 10 × faster than the GFK method and 20 × faster than the LTSL method.

4.4 Discussion

In this section, we discuss the results of the experiments. For action recognition by transferring missing knowledge from an existing source database, we did some experiments on the common RGB-D databases. MSR daily action 3D and MSR pair action 3D got results compared with the state-of-the-art transfer methods, LTSL and GFK. The experiments of different setting testified our previous expectation, and we present a discussion to follow.

First, our method can outperform other compared algorithms. In this chapter, we aim to improve the performance of recognition by transferring the depth information from the source domain to the target, using a latent low-rank framework. Test 1 and 2 were not traditional transfer learning cases, so that LTSL and GFK perform worse than the HOG-KNN method, as they would transfer more negative information from

another database. Another reason, we consider, is that those traditional transfer
learning methods do not consider the latent information and the cross-modality
structure, so they would fail to transfer the depth information; on the contrary, they
add more negative RGB information from the source data.

Second, we use two databases made up of different action categories. Therefore,
there is a large difference between two action categories, for example, pushing
a chair and sitting on the sofa, which are very different from the object/face
recognition. For instance, two faces from various people contain more similar
information, as the location of eyes and mouth can match respectively easily. So
it makes sense that the state-of-the-art transfer methods do not work well in the
situation of limited common knowledge. Notice the source database is related to the
target database to some extent; they should have similar visual contents, e.g., limbs
moving. Besides, we can see results from different cases and find that our method
performs well on the action recognition, regardless of the complex background,
meaning it is more robust than the other transfer methods that mainly rely on simple
experimental situation.

5 Summary

We addressed the problem by transferring depth information to the target RGB
action data (depth data is not available in the target data) and used both RGB
data and the learned depth data for action recognition. A novel transfer learning
method was proposed to use a RGB-D dataset as the source data and to learn a
shared subspace for RGB and depth data. The learned knowledge about the depth
information was then effectively transferred to the target data to enrich action
representations. Extensive experiments shows that our method achieved superior
performance over existing methods.

References

1. Cao, Y., Barrett, D., Barbu, A., Narayanaswamy, S., Yu, H., Michaux, A., Lin, Y., Dickinson, S., Siskind, J.M., Wang, S.: Recognize human activities from partially observed videos. In: CVPR, pp. 2658–2665. IEEE, New York (2013)
2. Choi, W., Shahid, K., Savarese, S.: Learning context for collective activity recognition. In: CVPR, pp. 3273–3280 (2011)
3. Ding, Z., Shao, M., Fu, Y.: Latent low-rank transfer subspace learning for missing modality recognition. In: AAAI (2014)
4. Dollar, P., Rabaud, V., Cottrell, G., Belongie, S.: Behavior recognition via sparse spatio-temporal features. In: VS-PETS, pp. 65–72 (2005)
5. Gong, B., Shi, Y., Sha, F., Grauman, K.: Geodesic flow kernel for unsupervised domain adaptation. In: CVPR, pp. 2066–2073 (2012)
6. Hadfield, S., Bowden, R.: Hollywood 3D: recognizing actions in 3D natural scenes. In: CVPR, pp. 3398–3405 (2013)

7. Hestenes, M.R.: Multiplier and gradient methods. J. Optim. Theory Appl. **4**(5), 303–320 (1969)
8. Jhuo, I.H., Liu, D., Lee, D., Chang, S.F.: Robust visual domain adaptation with low-rank reconstruction. In: CVPR, pp. 2168–2175 (2012)
9. Ji, S., Xu, W., Yang, M., Yu, K.: 3D convolutional neural networks for human action recognition. IEEE Trans. Pattern Anal. Mach. Intell. **35**(1), 221–231 (2013)
10. Jia, C., Zhong, G., Fu, Y.: Low-rank tensor learning with discriminant analysis for action classification and image recovery. In: AAAI (2014)
11. Klaser, A., Marszalek, M., Schmid, C.: A spatio-temporal descriptor based on 3d-gradients. In: BMVC, pp. 1–10 (2008)
12. Kolda, T.G., Bader, B.W.: Tensor decompositions and applications. SIAM Rev. **51**(3), 455–500 (2009)
13. Kong, Y., Jia, Y., Fu, Y.: Interactive phrases: semantic descriptions for human interaction recognition. IEEE Trans. Pattern Anal. Mach. Intell. **36**(9), 1775–1788 (2014)
14. Kong, Y., Kit, D., Fu, Y.: A discriminative model with multiple temporal scales for action prediction. In: ECCV (2014)
15. Laptev, I.: On space-time interest points. Int. J. Comput. Vis. **64**(2), 107–123 (2005)
16. Laptev, I., Marszałek, M., Schmid, C., Rozenfeld, B.: Learning realistic human actions from movies. In: CVPR, pp. 1–8 (2008)
17. Li, W., Vasconcelos, N.: Recognizing activities by attribute dynamics. In: NIPS, pp. 1106–1114 (2012)
18. Li, K., Fu, Y.: Prediction of human activity by discovering temporal sequence patterns. IEEE Trans. Pattern Anal. Mach. Intell. **36**(8), 1644–1657 (2014)
19. Li, X., Lin, S., Yan, S., Xu, D.: Discriminant locally linear embedding with high-order tensor data. IEEE Trans. Syst. Man Cybern. B **38**(2), 342–352 (2008)
20. Li, W., Zhang, Z., Liu, Z.: Action recognition based on a bag of 3d points. In: CVPR Workshop, pp. 9–14 (2010)
21. Liu, G., Yan, S.: Latent low-rank representation for subspace segmentation and feature extraction. In: ICCV, pp. 1615–1622 (2011)
22. Liu, L., Shao, L.: Learning discriminative representations from RGB-D video data. In: IJCAI, pp. 1493–1500 (2013)
23. Liu, J., Ali, S., Shah, M.: Recognizing human actions using multiple features. In: CVPR, pp. 1–8 (2008)
24. Liu, G., Lin, Z., Yu, Y.: Robust subspace segmentation by low-rank representation. In: ICML, pp. 663–670 (2010)
25. Liu, J., Kuipers, B., Savarese, S.: Recognizing human actions by attributes. In: CVPR, pp. 3337–3344 (2011)
26. Liu, G., Lin, Z., Yan, S., Sun, J., Yu, Y., Ma, Y.: Robust recovery of subspace structures by low-rank representation. IEEE Trans. Pattern Anal. Mach. Intell. **35**(1), 171–184 (2013)
27. Oreifej, O., Liu, Z.: HON4D: histogram of oriented 4D normals for activity recognition from depth sequences. In: CVPR, pp. 716–723 (2013)
28. Pan, S.J., Yang, Q.: A survey on transfer learning. IEEE Trans. Knowl. Data Eng. **22**(10), 1345–1359 (2010)
29. Pang, Y., Li, X., Yuan, Y.: Robust tensor analysis with l1-norm. IEEE Trans. Circuits Syst. Video Technol. **20**(2), 172–178 (2010)
30. Powell, M.J.D.: A method for nonlinear constraints in minimization problems. Optimization **60**(1), 283–298 (1969)
31. Raptis, M., Sigal, L.: Poselet key-framing: a model for human activity recognition. In: CVPR, pp. 2650–2657. IEEE, New York (2013)
32. Rodriguez, M.D., Ahmed, J., Shah, M.: Action mach: a spatio-temporal maximum average correlation height filter for action recognition. In: CVPR, pp. 1–8 (2008)
33. Ryoo, M., Aggarwal, J.: Spatio-temporal relationship match: video structure comparison for recognition of complex human activities. In: ICCV, pp. 1593–1600 (2009)

34. Ryoo, M.S., Aggarwal, J.K.: UT-interaction dataset, ICPR contest on semantic description of human activities (SDHA). In: IEEE International Conference on Pattern Recognition Workshops, vol. 2 (2010)
35. Schuldt, C., Laptev, I., Caputo, B.: Recognizing human actions: a local SVM approach. In: ICPR, pp. 32–36 (2004)
36. Shao, M., Castillo, C., Gu, Z., Fu, Y.: Low-rank transfer subspace learning. In: ICDM, pp. 1104–1109 (2012)
37. Shao, M., Kit, D., Fu, Y.: Generalized transfer subspace learning through low-rank constraint. Int. J. Comput. Vision 109(1–2), 74–93 (2014)
38. Soomro, K., Zamir, A.R., Shah, M.: Ucf101: a dataset of 101 human action classes from videos in the wild. In: CRCV-TR-12-01 (2013)
39. Sun, J., Tao, D., Papadimitriou, S., Yu, P.S., Faloutsos, C.: Incremental tensor analysis: theory and applications. Trans. Knowl. Discovery Data 2(3), 11 (2008)
40. Sutskever, I., Tenenbaum, J.B., Salakhutdinov, R.: Modelling relational data using Bayesian clustered tensor factorization. In: NIPS, pp. 1821–1828 (2009)
41. Tao, D., Song, M., Li, X., Shen, J., Sun, J., Wu, X., Faloutsos, C., Maybank, S.J.: Bayesian tensor approach for 3-d face modeling. IEEE Trans. Circuits Syst. Video Technol. 18(10), 1397–1410 (2008)
42. Tao, D., Li, X., Wu, X., Hu, W., Maybank, S.J.: Supervised tensor learning. Knowl. Inf. Syst. 13(1), 1–42 (2007)
43. Tao, D., Li, X., Wu, X., Maybank, S.J.: General tensor discriminant analysis and gabor features for gait recognition. IEEE Trans. Pattern Anal. Mach. Intell. 29(10), 1700–1715 (2007)
44. Von Luxburg, U.: A tutorial on spectral clustering. Stat. Comput. 17(4), 395–416 (2007)
45. Wang, Y., Mori, G.: Hidden part models for human action recognition: probabilistic vs. max-margin. IEEE Trans. Pattern Anal. Mach. Intell. 33(7), 1310–1323 (2011)
46. Wang, J., Liu, Z., Chorowski, J., Chen, Z., Wu, Y.: Robust 3D action recognition with random occupancy patterns. In: ECCV, pp. 872–885 (2012)
47. Wang, J., Liu, Z., Wu, Y., Yuan, J.: Mining actionlet ensemble for action recognition with depth cameras. In: CVPR, pp. 1290–1297 (2012)
48. Wang, Z., Wang, J., Xiao, J., Lin, K.H., Huang, T.S.: Substructural and boundary modeling for continuous action recognition. In: CVPR, pp. 1330–1337 (2012)
49. Xia, L., Aggarwal, J.: Spatio-temporal depth cuboid similarity feature for activity recognition using depth camera. In: CVPR, pp. 2834–2841 (2013)
50. Yan, S., Xu, D., Zhang, B., Zhang, H.J., Yang, Q., Lin, S.: Graph embedding and extensions: a general framework for dimensionality reduction. IEEE Trans. Pattern Anal. Mach. Intell. 29(1), 40–51 (2007)

Chapter 6
Activity Prediction

Yu Kong and Yun Fu

1 Introduction

Human action recognition [8, 11, 18, 19] is one of the active topics in the computer vision community, and has a broad range of applications, for example, video retrieval, visual surveillance, and video understanding.

After fully observing the entire video, action recognition approaches will classify the video observation into one of the action categories. It should be noted that certain real-world applications (e.g., vehicle accident and criminal activity) do not allow the luxury of waiting for the entire action to be executed. Reactions must be performed in a prompt to the action. For instance, it is extremely important to predict a dangerous driving situation before any vehicle crash occurs. Unfortunately, a majority of the existing action recognition approaches are limited to such particular scenarios since they must fully observe the action sequence extracted from the video.

One of the major differences between action prediction and action recognition is that action video data arrive sequentially in action prediction. However, action recognition takes the full observation as input. The key to perform early classification accurately is to extract the most discriminative information from the beginning segments in a temporal sequence. Furthermore, it is also important to

Y. Kong (✉)
Department of Electrical and Computer Engineering, Northeastern University,
360 Huntington Avenue, Boston, MA 02115, USA
e-mail: yukong@ece.neu.edu

Y. Fu
Department of Electrical and Computer Engineering and College of Computer and Information
Science (Affiliated), Northeastern University, 360 Huntington Avenue, Boston, MA 02115, USA
e-mail: yunfu@ece.neu.edu

© Springer International Publishing Switzerland 2016
Y. Fu (ed.), *Human Activity Recognition and Prediction*,
DOI 10.1007/978-3-319-27004-3_6

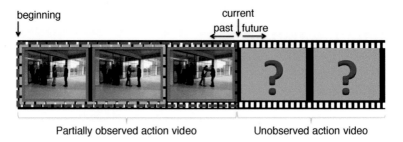

Fig. 6.1 Our method predicts action label given a partially observed video. Action dynamics are captured by both local templates (*solid rectangles*) and global templates (*dashed rectangles*)

effectively utilize history action information. The confidence of history observations is expected to increase since action data are progressively arriving in action prediction.

A novel multiple temporal scale support vector machine (MTSSVM) [9] is proposed in this chapter for early classification of unfinished actions. In MTSSVM, a human action is described at two different temporal granularities (Fig. 6.1). This allows us to learn the evolution and dynamics of actions, and make prediction from partially observed videos with temporally incomplete action executions. The sequential nature of human actions is considered at the fine granularity by local templates in the MTSSVM. The label consistency of temporal segments is enforced in order to maximize the discriminative information extracted from the segments. Note that the temporal orderings of inhomogeneous action segments is also captured by the temporal arrangement of these local templates in an implicit manner.

MTSSVM also capture history action information using coarse global templates. Different from local templates, the global templates characterize action evolutions at various temporal length, ranging from the beginning of the action video to the current frame. This global action information is effectively exploited in MTSSVM to differentiate between action categories. For instance, the key feature for differentiating action "push" from action "kick" is the motion "arm is up". Our model is learned for describing such increasing amount of information in order to capture featured motion evolution of each action class.

A new convex learning formulation is proposed in this chapter to consider the essence of the progressively arriving action data. The formulation is based on the structured SVM (SSVM), with new constraints being added. The label consistency is enforced between the full video and its containing temporal segments. This allows us to extract the discriminative information as much as possible for action prediction. Furthermore, a principled monotonic scoring function is modelled for the global templates. This scoring function enables us to utilize the fact that useful information is accumulating with the action data progressively arriving. We show that our new learning formulation can be efficiently solved using a standard SSVM solver. In addition, we demonstrate that the formulation essentially minimizes the upper bound of the empirical risk of the training data.

2   Related Work

Action Recognition   A popular representation for human actions is called bag-of-words approach, which characterizes the actions by a set of quantized local spatiotemporal features [3, 16, 18, 27]. Bag-of-words approach can capture local motion characteristics and insensitive to background noise. Nevertheless, it does not build expressive representation when large appearance and pose variations occur in videos. Researchers address this problem by integrating classification models with human knowledge and representing complex human actions by semantic descriptions or attributes [7, 8, 11]. Other solutions such as learning actions from a set of key frames [13, 23] or from status images [25, 26] have also been studied as well. Nevertheless, a majority of current action recognition algorithms are expected to fully observe actions before making predictions. This assumption hinders these algorithms from the task that human actions must be predicted when only partial of the action videos is observed.

Human actions can also be modeled as temporal evolutions of appearance or pose. This line of approaches generally utilize sequential state models [12, 20, 21, 24] to capture such evolutions, where a video is treated as an ordered temporal segments. However, the relationship of temporal action evolution in reference to observation ratios is not considered in these approaches, making them improper for action prediction. In comparison, the progressive data arrival is simulated in our approach. Large scale temporal templates are used to model action evolutions from the first frame to the current observed one. Hence, unfinished actions at various observation ratios can be accurately recognized using our approach.

Action Prediction   The goal of action prediction is to recognize unfinished action execution from partial videos. The integral bag-of-words (IBoW) and dynamic bag-of-words (DBoW) approaches were proposed in [15] for action prediction. These two approaches compute the mean of features in the same action category at the same progress level, and use the mean as the model for each progress level. Nevertheless, the constructed models are sensitive to outliers due to large intra-class appearance variations. This problem was overcome by [1], in which representative action models are built using the sparse coding technique. Results demonstrate that the proposed method achieves superior performance over the IBoW and DBoW approaches. All these method deal with short-duration action prediction problem, while long-duration problem was explored in [10]. One limitation of [10] is that the temporal segments are detected using motion velocity peaks that are very difficult to obtain in real-world outdoor datasets. Different from existing work [1, 10, 15], our prediction model integrates a crucial prior knowledge that the amount of useful information is accumulating with the arriving of new observations. This important prior information is not utilized in their methods. Furthermore, the proposed approach takes label consistency of segments into account, but it is not considered in their methods. Thanks to the label consistency, our approach is able to extract discriminative information in local segments and captures temporal ordering information implicitly. In addition, our model captures action dynamics at multiple scales while [1, 15] only capture the dynamics at one single scale.

Besides action prediction, [4] investigated early event detection problem. Their method can localize the beginning and ending frames given an unfinished event video. Kitani et al. [6] studied the problem of activity forecasting. The approach is able to reason the optimal path for a person to go from location A to location B.

3   Our Method

The aim of this work is to predict the action class y of a partially observed action video $x[1, t]$ before the action ends. Here 1 and t in $x[1, t]$ indicate the indices of the starting frame and the last observed frame of the partial video $x[1, t]$, respectively. Index t ranges from 1 to length T of a full video $x[1, T]$: $t \in \{1, \ldots, T\}$, to generate different partial videos. An action video is usually composed of a set of inhomogeneous temporal units, which are called segments. In this work, we uniformly divide a full video $x[1, T]$ into K segments $x[\frac{T}{K} \cdot (l - 1) + 1, \frac{T}{K} \cdot l]$, where $l = 1, \ldots, K$ is the index of segment. The length of each segment is $\frac{T}{K}$. Note that for different videos, their lengths T may be different. Therefore, the length of segments of various videos may be different. For simplicity, let $x_{(k)}$ be the kth segment $x[\frac{T}{K} \cdot (k-1) + 1, \frac{T}{K} \cdot k]$ and $x_{(1,k)}$ be the partially observed sequence $x[1, \frac{T}{K} \cdot k]$ (see Fig. 6.2). The progress level k of a partially observed video is defined as the number of observed segments that the video has. The observation ratio is the ratio of the number of frames in a partially observed video $x[1, t]$ to the number of frames in the full video $x[1, T]$, which is $\frac{t}{T}$. For example, if $T = 100$, $t = 30$ and $K = 10$, then the progress level of the partially observed video $x[1, t]$ is 3 and its observation ratio is 0.3.

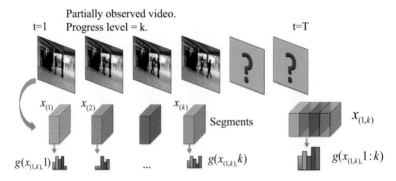

Fig. 6.2 Example of video segments $x_{(k)}$, partial video $x_{(1,k)}$, feature representations $g(x_{(1,k)}, l)$ of segments ($l = 1, \ldots, k$), and the representation of the partial video $g(x_{(1,k)}, 1 : k)$

3.1 Action Representations

We use the bag-of-words models to represent segments and partial videos. The procedure of learning the visual word dictionary for action videos is as follows. Spatiotemporal interest points detector [3] and tracklet [14] are employed to extract interest points and trajectories from a video, respectively. The dictionaries of visual words are learned by clustering algorithms.

We denote the feature of the partial video $x_{(1,k)}$ at progress level k by $g(x_{(1,k)}, 1:k)$, which is the histogram of visual words contained in the entire partial video, starting from the first segment to the kth segment (Fig. 6.2). The representation of the lth $(l \in \{1, \ldots, k\})$ segment $x_{(l)}$ in the partial video is denoted by $g(x_{(1,k)}, l)$, which is a histogram of visual words whose temporal locations are within the lth segment.

3.2 Model Formulation

Let $\mathcal{D} = \{x_i, y_i\}_{i=1}^N$ be the training data, where x_i is the ith fully observed action video and y_i is the corresponding action label. The problem of action prediction is to learn a function $f : \mathcal{X} \to \mathcal{Y}$, which maps a partially observed video $x_{(1,k)} \in \mathcal{X}$ to an action label $y \in \mathcal{Y}$ ($k \in \{1, \ldots, K\}$).

We formulate the action prediction problem using the structured learning as presented in [22]. Instead of searching for f, we aim at learning a discriminant function $F : \mathcal{X} \times \mathcal{Y} \to \mathcal{R}$ to score each training sample (x, y). The score measures the compatibility between a video x and an action label y. Note that, in action prediction, videos of different observation ratios from the same class should be classified as the same action category. Therefore, we use the function F to score the compatibility between the videos of different observation ratios $x_{(1,k)}$ and the action label y, where $k \in \{1, \ldots, K\}$ is the progress level.

We are interested in a linear function $F(x_{(1,k)}, y; \mathbf{w}) = \langle \mathbf{w}, \Phi(x_{(1,k)}, y) \rangle$, which is a family of functions parameterized by $\mathbf{w}$, and $\Phi(x_{(1,k)}, y)$ is a joint feature map that represents the spatio-temporal features of action label y given a partial video $x_{(1,k)}$. Once the optimal model parameter $\mathbf{w}^*$ is learned, the prediction of the action label y^* is computed by

$$y^* = \arg\max_{y \in \mathcal{Y}} F(x_{(1,k)}, y; \mathbf{w}^*) = \arg\max_{y \in \mathcal{Y}} \langle \mathbf{w}^*, \Phi(x_{(1,k)}, y) \rangle. \tag{6.1}$$

We define $\mathbf{w}^T \Phi(x_{(1,k)}, y)$ as a summation of the following two components:

$$\mathbf{w}^T \Phi(x_{(1,k)}, y) = \boldsymbol{\alpha}_k^T \psi_1(x_{(1,k)}, y) + \sum_{l=1}^{K} \left[\mathbf{1}(l \le k) \cdot \boldsymbol{\beta}_l^T \psi_2(x_{(1,k)}, y) \right], \tag{6.2}$$

where $\mathbf{w} = \{\alpha_1, \ldots \alpha_K, \beta_1, \ldots, \beta_K\}$ is model parameter, k is the progress level of the partial video $x_{(1,k)}$, l is the index of progress levels, and $\mathbf{1}(\cdot)$ is the indicator function. The two components in Eq. (6.2) are summarized as follows.

Global progress model (GPM) $\alpha_k^T \psi_1(x_{(1,k)}, y)$ indicates how likely the action class of an unfinished action video $x_{(1,k)}$ (at progress level k) is y. We define GPM as

$$\alpha_k^T \psi_1(x_{(1,k)}, y) = \sum_{a \in \mathcal{Y}} \alpha_k^T \mathbf{1}(y = a) g(x_{(1,k)}, 1 : k). \quad (6.3)$$

Here, feature vector $g(x_{(1,k)}, 1 : k)$ of dimensionality D is an action representation for the partial video $x_{(1,k)}$, where features are extracted from the entire partial video, from its beginning (i.e., progress level 1) to its current progress level k. Parameter α_k of size $D \times |\mathcal{Y}|$ can be regarded as a progress level-specific template. Since the partial video is at progress level k, we select the template α_k at the same progress level, from K parameter matrices $\{\alpha_1, \ldots, \alpha_K\}$. The selected template α_k is used to score the unfinished video $x_{(1,k)}$. Define $A = [\alpha_1, \ldots, \alpha_K]$ as a vector of all the parameter matrices in the GPM. Then A is a vector of size $D \times K \times |\mathcal{Y}|$ encoding the weights for the configurations between progress levels and action labels, with their corresponding video evidence.

The GPM simulates the sequential segment-by-segment data arrival for training action videos. Essentially, the GPM captures the action appearance changes as the progress level increases, and characterizes the entire action evolution over time. In contrast to the IBoW model [15], our GPM does not assume any distributions on the data likelihood; while the IBoW model uses the Gaussian distribution. In addition, the compatibility between observation and action label in our model is given by the linear model of parameter and feature function, rather than using a Gaussian kernel function [15].

Local progress model (LPM) $\mathbf{1}(l \leq k) \cdot \beta_l^T \psi_2(x_{(1,k)}, y)$ indicates how likely the action classes of all the temporal segments $x_{(l)}$ ($l = 1, \ldots, k$) in an unfinished video $x_{(1,k)}$ are all y. Here, the progress level of the partial video is k and we consider all the segments of the video whose temporal locations l are smaller than k. We define LPM as

$$\beta_l^T \psi_2(x_{(1,k)}, y) = \sum_{a \in \mathcal{Y}} \beta_l^T \mathbf{1}(y = a) g(x_{(1,k)}, l), \quad (6.4)$$

where feature vector $g(x_{(1,k)}, l)$ of dimensionality D extracts features from the lth segment of the unfinished video $x_{(1,k)}$. β_l of size $D \times |\mathcal{Y}|$ is the weight matrix for the lth segment. We use the indicator function $\mathbf{1}(l \leq k)$ to select all the segment weight matrices, $\beta_1, \ldots, \beta_k$, whose temporal locations are smaller than or equal to the progress level k of the video. Then the selected weight matrices are used to score the corresponding segments. Let $B = [\beta_1, \ldots, \beta_K]$ be a vector of all the parameters in the LPM. Then B is a vector of size $D \times K \times |\mathcal{Y}|$ encoding the weights for the configurations between segments and action labels, with their corresponding segment evidence.

The LPM considers the sequential nature of a video. The model decomposes a video of progress level k into segments and describes the temporal dynamics of segments. Note that the action data preserve the temporal relationship between the segments. Therefore, the discriminative power of segment $x_{(k)}$ is critical to the prediction of $x_{(1,k)}$ given the prediction results of $x_{(1,k-1)}$. In this work, the segment score $\beta_k^T g(x_{(1,k)}, k)$ measures the compatibility between the segment $x_{(k)}$ and all the classes. To maximize the discriminability of the segment, the score difference between the ground-truth class and all the other classes is maximized in our learning formulation. Thus, accurate prediction can be achieved using the newly introduced discriminative information in the segment $x_{(k)}$.

3.3 Structured Learning Formulation

The MTSSVM is formulated based on the SSVM [5, 22]. The optimal model parameter $\mathbf{w}^*$ of MTSSVM in Eq. (6.1) is learned by solving the following convex problem given training data $\{x_i, y_i\}_{i=1}^N$:

$$\min \frac{1}{2}\|\mathbf{w}\|^2 + \frac{C}{N}\sum_{i=1}^N (\xi_{1i} + \xi_{2i} + \xi_{3i}) \tag{6.5}$$

$$\text{s.t. } \mathbf{w}^T \Phi(x_{i(1,k)}, y_i) \geq \mathbf{w}^T \Phi(x_{i(1,k)}, y) + K\delta(y, y_i) - \frac{\xi_{1i}}{u(k/K)}, \quad \forall i, \forall k, \forall y, \tag{6.6}$$

$$\alpha_k^T \psi_1(x_{i(1,k)}, y_i) \geq \alpha_{k-1}^T \psi_1(x_{i(1,k-1)}, y) + K\delta(y, y_i) - \frac{\xi_{2i}}{u(k/K)},$$

$$\forall i, k = 2, \ldots, K, \forall y, \tag{6.7}$$

$$\beta_k^T \psi_2(x_{i(k)}, y_i) \geq \beta_k^T \psi_2(x_{i(k)}, y) + kK\delta(y, y_i) - \frac{\xi_{3i}}{u(1/K)}, \quad \forall i, \forall k, \forall y, \tag{6.8}$$

where C is the slack trade-off parameter similar to that in SVM. ξ_{1i}, ξ_{2i}, and ξ_{3i} are slack variables. $u(\cdot)$ is a scaling factor function: $u(p) = p$. $\delta(y, y_i)$ is the 0-1 loss function.

The slack variables ξ_{1i} and the Constraint (6.6) are usually used in SVM constraints on the class labels. We enforce this constraint for all the progress levels k since we are interested in learning a classifier that can correctly recognize partially observed videos with different progress levels k. Therefore, we simulate the segment-by-segment data arrival for training and augment the training data with partial videos of different progress levels. The loss function $\delta(y, y_i)$ measures the recognition error of a partial video and the scaling factor $u(\frac{k}{K})$ scales the loss based on the length of the partial video.

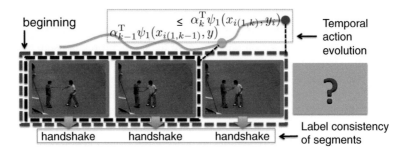

Fig. 6.3 Graphical illustration of the temporal action evolution over time and the label consistency of segments. *Blue solid rectangles* are LPMs, and *purple* and *red dashed rectangles* are GPMs

Constraint (6.7) considers **temporal action evolution** over time (Fig. 6.3). We assume that the score $\boldsymbol{\alpha}^T \psi_1(x_{i(1,k)}, y_i)$ of the partial observation $x_{i(1,k)}$ at progress level k and ground truth label y_i must be greater than the score $\boldsymbol{\alpha}^T \psi_1(x_{i(1,k-1)}, y)$ of a previous observation $x_{i(1,k-1)}$ at progress level $k - 1$ and all incorrect labels y. This provides a monotonically increasing score function for partial observations and elaborately characterizes the nature of sequentially arriving action data in action prediction. The slack variable ξ_{2i} allows us to model outliers.

The slack variables ξ_{3i} and the Constraint (6.8) are used to maximize the discriminability of segments $x_{(k)}$. We encourage the **label consistency** between segments and the corresponding full video due to the nature of sequential data in action prediction (Fig. 6.3). Assume a partial video $x_{(1,k-1)}$ has been correctly recognized, then the segment $x_{(k)}$ is the only newly introduced information and its discriminative power is the key to recognizing the video $x_{(1,k)}$. Moreover, context information of segments is implicitly captured by enforcing the label consistency. It is possible that some segments from different classes are visually similar and may not be linearly separable. We use the slack variable ξ_{3i} for each video to allow some segments of a video to be treated as outliers.

Empirical Risk Minimization We define $\Delta(y_i, y)$ as the function that quantifies the loss for a prediction y, if the ground-truth is y_i. Therefore, the loss of a classifier $f(\cdot)$ for action prediction on a video-label pair (x_i, y_i) can be quantified as $\Delta(y_i, f(x_i))$. Usually, the performance of $f(\cdot)$ is given by the empirical risk $R_{\text{emp}}(f) = \frac{1}{N} \sum_{i=1}^{N} \Delta(y_i, f(x_i))$ on the training data (x_i, y_i), assuming data samples are generated i.i.d.

The nature of continual evaluation in action prediction requires aggregating the values of loss quantities computed during the action sequence process. Define the loss associated with a prediction $y = f(x_{i(1,k)})$ for an action x_i at progress level k as $\Delta(y_i, y)u(\frac{k}{K})$. Here $\Delta(y_i, y)$ denotes the misclassification error, and $u(\frac{k}{K})$ is the scaling factor that depends on how many segments have been observed. In this work, we use summation to aggregate the loss quantities. This leads to an empirical risk for N training samples: $R_{\text{emp}}(f) = \frac{1}{N} \sum_{i=1}^{N} \sum_{k=1}^{K} \{\Delta(y_i, y)u(\frac{k}{K})\}$.

Denote by $\boldsymbol{\xi}_1^*, \boldsymbol{\xi}_2^*$ and $\boldsymbol{\xi}_3^*$ the optimal solutions of the slack variables in Eq. (6.5)–(6.8) for a given classifier f, we can prove that $\frac{1}{N} \sum_{i=1}^{N} (\xi_{1i}^* + \xi_{2i}^* + \xi_{3i}^*)$ is an upper bound on the empirical risk $R_{\mathrm{emp}}(f)$ and the learning formulation given in Eq. (6.5)–(6.8) minimizes the upper bound of the empirical risk $R_{\mathrm{emp}}(f)$.

3.4 Discussion

We highlight here some important properties of our model, and show some differences from existing methods.

Multiple Temporal Scales Our method captures action dynamics in both local and global temporal scales, while [1, 4, 15] only use a single temporal scale.

Temporal Evolution Over Time Our work uses the prior knowledge of temporal action evolution over time. Inspired by [4], we introduce a principled monotonic score function for the GPM to capture this prior knowledge. However, [4] aims at finding the starting frame of an event while our goal is to predict action class of an unfinished video. The methods in [1, 10, 15] do not use this prior.

Segment Label Consistency We effectively utilize the discriminative power of local temporal segments by enforcing label consistency of segments. However, [1, 4, 10, 15] do not consider the label consistency. The consistency also implicitly models temporal segment context by enforcing the same label for segments while [1, 4, 15] explicitly treat successive temporal segments independently.

Principled Empirical Risk Minimization We propose a principled empirical risk minimization formulation for action prediction, which is not discussed in [1, 10, 15].

3.5 Model Learning and Testing

Learning We solve the optimization problem (6.5)–(6.8) using the regularized bundle algorithm [2]. The basic idea of the algorithm is to iteratively approximate the objective function by adding a new cutting plane to the piecewise quadratic approximation.

The equivalent unconstrained problem of the optimization problem (6.5)–(6.8) is $\min_{\mathbf{w}} \frac{1}{2} \|\mathbf{w}\|^2 + \frac{C}{N} \cdot L(\mathbf{w})$, where $L(\mathbf{w}) = \sum_{i=1}^{N} (U_i + Z_i + V_i)$ is the empirical loss. Here, U_i, Z_i, and V_i are given by

$$U_i = \sum_{k=1}^{K} u\left(\frac{k}{K}\right) \max_{y} \left[K\delta(y, y_i) + \mathbf{w}^{\mathsf{T}} \Phi(x_{i(1,k)}, y) - \mathbf{w}^{\mathsf{T}} \Phi(x_{i(1,k)}, y_i) \right], \qquad (6.9)$$

$$Z_i = \sum_{k=2}^{K} u\left(\frac{k}{K}\right) \max_y \left[K\delta(y, y_i) + \boldsymbol{\alpha}_{k-1}^{\mathrm{T}} \psi_1(x_{i(1,k-1)}, y) - \boldsymbol{\alpha}_k^{\mathrm{T}} \psi_1(x_{i(1,k)}, y_i) \right],$$

$$(6.10)$$

$$V_i = \sum_{k=1}^{K} u\left(\frac{1}{K}\right) \max_y \left[kK\delta(y, y_i) + \boldsymbol{\beta}_k^{\mathrm{T}} \psi_2(x_{i(k)}, y) - \boldsymbol{\beta}_k^{\mathrm{T}} \psi_2(x_{i(k)}, y_i) \right]. \qquad (6.11)$$

The regularized bundle algorithm requires the subgradient of the training loss with respect to the parameter, $\frac{\partial L}{\partial \mathbf{w}} = \sum_{i=1}^{N}(\frac{\partial U_i}{\partial \mathbf{w}} + \frac{\partial Z_i}{\partial \mathbf{w}} + \frac{\partial V_i}{\partial \mathbf{w}})$, in order to find a new cutting plane to be added to the approximation.

Testing Given an unfinished action video with progress level k (k is known in testing), our goal is to infer the class label y^* using the learned model parameter $\mathbf{w}^*$: $y^* = \arg\max_{y \in \mathcal{Y}} \langle \mathbf{w}^*, \Phi(x_{(1,k)}, y) \rangle$. Note that testing phase does not require sophisticated inference algorithms such as belief propagation or graph cut since we do not explicitly capture segment interactions. However, the context information between segments is implicitly captured in our model by the label consistency in Constraint (6.8).

4 Experiments

We test the proposed MTSSVM approach on three datasets: the UT-Interaction dataset (UTI) Set 1 (UTI #1) and Set 2 (UTI #2) [17], and the BIT-Interaction dataset (BIT) [7]. UTI #1 were taken on a parking lot with mostly static background and little camera jitters. UTI #2 were captured on a lawn with slight background movements (e.g., tree moves) and camera jitters. Both of the two sets consist of six types of human actions, with ten videos per class. We adopt the leave-one-out training scheme on the two datasets. The BIT dataset consists of eight types of human actions between two people, with 50 videos per class. For this dataset, a random sample of 272 videos is chosen as training samples, and the remaining 128 videos are used for testing. The dictionary size for interest point descriptors is set to 500, and the size for tracklet descriptors is automatically determined by the clustering method in all the experiments.

MTSSVM is evaluated for classifying videos of incomplete action executions using ten observation ratios, from 0.1 to 1, representing the increasing amount of sequential data with time. For example, if a full video containing T frames is used for testing at the observation ratio of 0.3, the accuracy of MTSSVM is evaluated by presenting it with the first $0.3 \times T$ frames. At observation ratio of 1, the entire video is used, at which point MTSSVM acts as a conventional action recognition model. The progress level k of testing videos is known to all the methods in our experiments.

4.1   Results

UTI #1 and UTI #2 Datasets The MTSSVM is compared with DBoW and IBoW in [15], the MMED [4], the MSSC and the SC in [1], and the method in [13]. The KNN-nonDynamic, the KNN-Dynamic, and the baseline method implemented in [1] are also used in comparison. The same experiment settings in [1] are followed in our experiments.

Figure 6.4a shows the prediction results on the UTI #1 dataset. Our MTSSVM achieves better performance over all the other comparison approaches. Our method outperforms the MSSC method because we not only model segment dynamics but also characterize temporal evolutions of actions. Our method can achieve an impressive 78.33 % recognition accuracy when only the first 50 % frames of testing videos are observed. This result is even higher than the SC method with full observations. Results of our method are significantly higher than the DBoW and IBoW for all observation ratios. This is mainly due to the fact that the action models in our work are discriminatively learned while the action models in the DBoW and IBoW are computed by averaging feature vectors in a particular class. Therefore, the action models in the DBoW and IBoW may not be the representative models and are sensitive to outliers. MMED does not perform well as other prediction approaches since it is optimized for early detection of the starting and ending frame of an action. This is a different goal from this chapter, which is to classify unfinished actions. We also compare with [13] on half and full video observations. Results in Table 6.1 show that our method achieves better performance over [13].

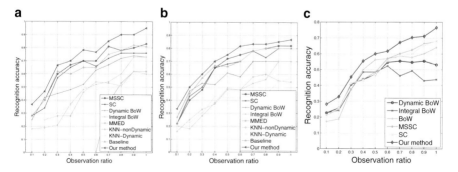

Fig. 6.4 Prediction results on the (**a**) UTI #1, (**b**) UTI #2, and (**c**) BIT datasets

Table 6.1 Prediction results compared with [13] on half and full videos

Observation ratio	Accuracy with half videos (%)	Accuracy with full videos (%)
Raptis and Sigal [13]	73.3	93.3
Our model	**78.33**	**95**

Comparison results on the UTI #2 datasets are shown in Fig. 6.4b. The MTSSVM achieves better performance over all the other comparison approaches in all the cases. At 0.3, 0.5, and 1 observation ratios, MSSC achieves 48.33 %, 71.67 %, and 81.67 % prediction accuracy, respectively, and SC achieves 50 %, 66.67 %, and 80 % accuracy, respectively. By contrast, our MTSSVM achieves 60 %, 75%, and 83.33 % prediction results, respectively, which is consistently higher than MSSC and SC. Our MTSSVM achieves 75 % accuracy when only the first 50 % frames of testing videos are observed. This accuracy is even higher than the DBoW and IBoW with full observations.

To demonstrate that both the GPM and the LPM are important for action prediction, we compare the performance of MTSSVM with the model that only uses one of the two sources of information on the UTI #1 dataset. Figure 6.5 shows the scores of the GPM and LPM ($\alpha_k^T \psi_1(x_{(1,k)}, y)$ of the GPM and $\sum_{l=1}^{K} \mathbf{1}(l \leq k) \cdot \beta_l^T \psi_2(x_{(1,k)}, y)$ of the LPM), and compare them to the scores of the full MTSSVM model with respect to the observation ratio. Results show that the LPM captures discriminative temporal segments for prediction. LPM characterizes temporal dynamics of segments and discriminatively learns to differentiate segments from different classes. In most cases, the score of LPM is monotonically increasing, which indicates a discriminative temporal segment is used for prediction. However, in some cases, segments from different classes are visually similar and thus are difficult to discriminate. Therefore, in the middle of the "handshake" class and the "hug" class in Fig. 6.5 (observation ratio from 0.3 to 0.7), adding more segment observations does not increase LPM's contribution to MTSSVM. Figure 6.6 shows examples of visually similar segments of the two classes at $k = 6$. However, when such situations arise, GPM can provide necessary appearance history information and therefore increases the prediction performance of MTSSVM.

BIT-Interaction Dataset We also compare MTSSVM with the MSSC, SC, DBoW and IBoW on the BIT-Interaction dataset. A BoW+SVM method is used as a baseline. The parameter σ in DBoW and IBoW is set to 36 and 2, respectively, which are the optimal parameters on the BIT-Interaction dataset. Results shown in Fig. 6.4c demonstrate that MTSSVM outperforms MSSC and SC in all cases due to the effect of the GPM, which effectively captures temporal action evolution information. MTSSVM also outperforms the DBoW and IBoW. Our method achieves 60.16 % recognition accuracy with only the first 50 % frames of testing videos are observed, which is better than the DBoW and IBoW at all observation ratios. Note that the performance of DBoW and IBoW does not increase much when the observation ratios are increased from 0.6 to 0.9. The IBoW performs even worse. This is due to the fact that some video segments from different classes are visually similar;

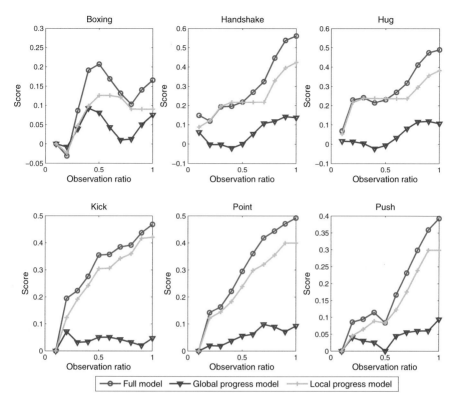

Fig. 6.5 Contributions of the global progress model and the local progress model to the prediction task

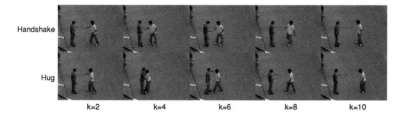

Fig. 6.6 Examples of segments in "handshake" and "hug". Segments $k = 6, 8, 10$ in the two classes are visually similar

especially, the segments in the second half of the videos, where people return to their starting positions (see Fig. 6.7). However, because MTSSVM models both the segments and the entire observation, its performance increases with the increasing of observation ratio even if the newly introduced segments contain only a small amount of discriminative information.

We further investigate the sensitivity of MTSSVM to the parameters C in Eq. (6.5). We set C to 0.5, 5, and 10, and test MTSSVM on all parameter

Fig. 6.7 Examples of visually similar segments in the "boxing" action (*Top*) and the "pushing" action (*Bottom*) with segment index $k \in \{2, 4, 6, 8, 10\}$. *Bounding boxes* indicate the interest regions of actions

Table 6.2 Recognition accuracy of our model on videos of observation ratio 0.3, 0.5, and 0.8 with different C parameters

Observation ratio	$C = 0.5$ (%)	$C = 5$ (%)	$C = 10$ (%)
0.3	42.97	39.84	38.28
0.5	54.69	57.03	51.56
0.8	66.41	61.72	55.47

combinations with observation ratios 0.3, 0.5, and 0.8. Results in Table 6.2 indicate that MTSSVM is not sensitive to the parameters when the observation ratio is low but the sensitivity increases when the observation ratio becomes large. In the beginning of a video, the small number of features available does not capture the variability of their class. Therefore, it does not help to use different parameters, because MTSSVM cannot learn the appropriate class boundaries to separate all the testing data. As observation ratio increases, the features become more expressive. However, since structural features in MTSSVM are very complex, appropriate parameters are required to capture the complexity of data.

Finally, we also evaluate the importance of each component in the MTSSVM, including the Constraint (6.7), the Constraint (6.8), the local progress model [LPM in Eq. (6.4)], and the global progress model [GPM in Eq. (6.3)]. We remove each of these components from the MTSSVM, and obtain four variant models, the no-cons2 model [remove the Constraint (6.7) from MTSSVM], the no-cons3 model [remove the Constraint (6.8)], the no-LPM model [remove the LPM and Constraint (6.8)], and the no-GPM model [remove the GPM and Constraint (6.7)]. We compare MTSSVM with these variants with parameter C of 1 and 100. Results in Fig. 6.8 show that the GPM is the key component in the MTSSVM. Without the GPM, the performance of the no-GPM model degrades significantly compared with the full MTSSVM model, especially with parameter C of 100. The performances of the no-cons3 model and the no-LPM model are worse compared with the full method in all cases. This is due to the lack of the segment label consistency in the two models. The label consistency can help use the discriminative information in segments and also implicitly model context information. In the ending part of videos in BIT dataset, since most of the observations are visually similar (people return back to their normal position), label consistency is of great importance for discriminating classes. However, due to the lack of label consistency in the no-cons3 model and the no-LPM model, they cannot capture useful information for differentiating action classes.

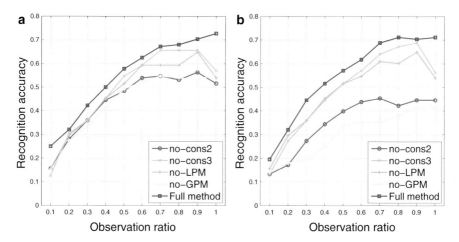

Fig. 6.8 Prediction results of each component in the full MTSSVM with *C* parameter (**a**) 1 and (**b**) 100

5   Summary

We have proposed the MTSSVM for recognizing actions in incomplete videos. MTSSVM captures the entire action evolution over time and also considers the temporal nature of a video. We formulate the action prediction task as a SSVM learning problem. The discriminability of segments is enforced in the learning formulation. Experiments on two datasets show that MTSSVM outperforms state-of-the-art approaches.

References

1. Cao, Y., Barrett, D., Barbu, A., Narayanaswamy, S., Yu, H., Michaux, A., Lin, Y., Dickinson, S., Siskind, J., Wang, S.: Recognizing human activities from partially observed videos. In: CVPR (2013)
2. Do, T.-M.-T., Artieres, T.: Large margin training for hidden Markov models with partially observed states. In: ICML (2009)
3. Dollar, P., Rabaud, V., Cottrell, G., Belongie, S.: Behavior recognition via sparse spatio-temporal features. In: VS-PETS (2005)
4. Hoai, M., De la Torre, F.: Max-margin early event detectors. In: CVPR (2012)
5. Joachims, T., Finley, T., Yu, C.-N.: Cutting-plane training of structural SVMs. Mach. Learn. **77**(1), 27–59 (2009)
6. Kitani, K.M., Ziebart, B.D., Andrew Bagnell, J., Martial Hebert, M.: Activity forecasting. In: ECCV (2012)
7. Kong, Y., Jia, Y., Fu, Y.: Learning human interaction by interactive phrases. In: ECCV (2012)
8. Kong, Y., Jia, Y., Fu, Y.: Interactive phrases: semantic descriptions for human interaction recognition. In: TPAMI (2014)

9. Kong, Y., Kit, D., Fu, Y.: A discriminative model with multiple temporal scales for action prediction. In: ECCV (2014)
10. Li, K., Hu, J., Fu, Y.: Modeling complex temporal composition of actionlets for activity prediction. In: ECCV (2012)
11. Liu, J., Kuipers, B., Savarese, S.: Recognizing human actions by attributes. In: CVPR (2011)
12. Niebles, J.C., Chen, C.-W., Fei-Fei, L.: Modeling temporal structure of decomposable motion segments for activity classification. In: ECCV (2010)
13. Raptis, M., Sigal, L.: Poselet key-framing: a model for human activity recognition. In: CVPR (2013)
14. Raptis, M., Soatto, S.: Tracklet descriptors for action modeling and video analysis. In: ECCV (2010)
15. Ryoo, M.S.: Human activity prediction: early recognition of ongoing activities from streaming videos. In: ICCV (2011)
16. Ryoo, M.S., Aggarwal, J.K.: Spatio-temporal relationship match: video structure comparison for recognition of complex human activities. In: ICCV, pp. 1593–1600 (2009)
17. Ryoo, M., Aggarwal, J.: UT-interaction dataset, ICPR contest on semantic description of human activities. http://cvrc.ece.utexas.edu/SDHA2010/Human_Interaction.html (2010)
18. Schuldt, C., Laptev, I., Caputo, B.: Recognizing human actions: a local SVM approach. In: ICPR, vol. 3, pp. 32–36. IEEE, New York (2004)
19. Shapovalova, N., Vahdat, A., Cannons, K., Lan, T., Mori, G.: Similarity constrained latent support vector machine: an application to weakly supervised action classification. In: ECCV (2012)
20. Shi, Q., Cheng, L., Wang, L., Smola, A.: Human action segmentation and recognition using discriminative semi-Markov models. Int. J. Comput. Vis. **93**, 22–32 (2011)
21. Tang, K., Fei-Fei, L., Koller, D.: Learning latent temporal structure for complex event detection. In: CVPR (2012)
22. Tsochantaridis, I., Joachims, T., Hofmann, T., Altun, Y.: Large margin methods for structured and interdependent output variables. J. Mach. Learn. Res. **6**, 1453–1484 (2005)
23. Vahdat, A., Gao, B., Ranjbar, M., Mori, G.: A discriminative key pose sequence model for recognizing human interactions. In: ICCV Workshops, pp. 1729–1736 (2011)
24. Wang, Z., Wang, J., Xiao, J., Lin, K.-H., Huang, T.S.: Substructural and boundary modeling for continuous action recognition. In: CVPR (2012)
25. Yao, B., Fei-Fei, L.: Action recognition with exemplar based 2.5d graph matching. In: ECCV (2012)
26. Yao, B., Fei-Fei, L.: Recognizing human-object interactions in still images by modeling the mutual context of objects and human poses. IEEE Trans. Pattern Anal. Mach. Intell. **34**(9), 1691–1703 (2012)
27. Yu, T.-H., Kim, T.-K., Cipolla, R.: Real-time action recognition by spatiotemporal semantic and structural forests. In: BMVC (2010)

Chapter 7
Actionlets and Activity Prediction

Kang Li and Yun Fu

1 Introduction

The increasing ubiquitousness of multimedia information in today's world has positioned video as a favored information vehicle, and given rise to an astonishing generation of social media and surveillance footage. One important problem that will significantly enhance semantic-level video analysis is activity understanding, which aims at accurately describing video contents using key semantic elements, especially activities. We notice that in case a time-critical decision is needed, there is a potential to utilize the temporal structure of videos for early prediction of ongoing human activity.

In recent years, research shows that modeling temporal structure is a basic methodology for recognition of complex human activity [6, 20, 42]. These studies extend the types of human activity that can be understood by machine vision systems. Advances in this field made an important application become real: *predicting activities or imminent events from observed actions or events in the video.*

ⓒ {Kang Li and Yun Fu | IEEE}, {2014}. This is a minor revision of the work published in {Pattern Analysis and Machine Intelligence, IEEE Transactions on, pp. 1644–1657. vol.36, no.8.}, http://dx.doi.org/10.1109/TPAMI.2013.2297321.

K. Li (✉)
Department of Electrical and Computer Engineering, Northeastern University,
360 Huntington Avenue, Boston, MA 02115, USA
e-mail: li.ka@husky.neu.edu; kongkong115@gmail.com

Y. Fu
Department of Electrical and Computer Engineering and College of Computer and Information Science (Affiliated), Northeastern University, 360 Huntington Avenue, Boston, MA 02115, USA
e-mail: yunfu@ece.neu.edu

© Springer International Publishing Switzerland 2016
Y. Fu (ed.), *Human Activity Recognition and Prediction*,
DOI 10.1007/978-3-319-27004-3_7

Many intelligence systems can benefit from activity prediction. For instance, in the sports video analysis, the capability of predicting the progress or results of a sports game is highly desirable. In public areas, we want to equip a surveillance system that can raise an alarm in advance of any potential dangerous activity happens. In a smart room, people's intention of activity can be predicted by a user-friendly sensor—camera, so that the system will adaptively provide services, even help if necessary.

Though activity prediction is a very interesting and important problem, it is quite a new topic for the domain of computer vision. One well-known challenge is the long-standing semantic gap between computable low-level features and semantic information that they encode. To the best of our knowledge, the work in [46] is the only one that explicitly focused on this problem. They identified activity prediction with early detection of short-duration single action, such as "hugging," "pushing." This assumption limits the types of activities that can be predicted as well as how early the prediction can be made. We believe that activity prediction is more desirable and valuable if it focuses on long-duration complex activities, such as "making a sandwich." The early detection problem can be solved in the classic recognition paradigm by predicting directly on low-level feature representations. Our approach aims to solve the long duration prediction problem with a completely different framework, where semantic-level understanding and reasoning are our focus.

Specifically, in this chapter, we present a novel approach for predicting long-duration complex activity by discovering the causal relationships between constituent actions and predictable characteristic of the activities. The key of our approach is to utilize the observed action units as context to predict the next possible action unit, or predict the intension and effect of the whole activity. It is thus possible to make predictions with meaningful earliness and have the machine vision system provide a time-critical reaction. We represent complex activities as sequences of discrete action units, which have specific semantic meanings and clear time boundaries. To ensure a good discretization, we propose a novel temporal segmentation method for action units by discovering the regularity of motion velocities. We argue that the causality of action units can be encoded as Markov dependencies with various lengths, while the predictability can be characterized by a predictive accumulative function (PAF) learned from information entropy changes along every stage of activity progress.

Additionally, according to cognitive science, context information is critical for understanding human activities [10, 21, 23, 25, 28, 35, 55], which typically occur under particular scene settings with certain object interactions. So for activity prediction, it needs to involve not only actions, but also objects and their spatial temporal arrangement with actions. Such knowledge can provide valuable clues for two questions *'what is happening now?'* and *'what is goanna happen next?'*. Therefore, a unified approach is expected to provide unexplored opportunities to benefit from mutual contextual constraints among actions and objects. When a particular ⟨*action, object*⟩ pair is observed, the whole plan of human behavior may be inferred immediately. For example, as long as we observe *'a person grabbing*

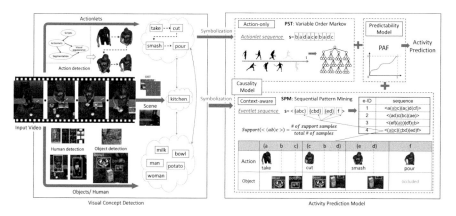

Fig. 7.1 Frameworks of long-duration complex activity prediction. Two scenarios: (1) Action-only activity prediction. (2) Context-aware activity prediction. The particular activity shown in the sample video is "making mashed potatoes." The video data is from [44]. Our approach aims to solve activity prediction problem in both cases

a cup,' we probably can tell s/he is going to drink a beverage. In this chapter, we will introduce both an action-only model [33] and a context-aware model [32]. We utilize sequential pattern mining (SPM) to incorporate the context information into actions which together can be represented as enriched symbolic sequences.

Overall, we propose a generalized activity prediction framework, which has four major components as shown in Fig. 7.1: (1) a visual concept detection module, (2) an action-only causality model, (3) a context-aware causality model, and (4) a predictability model. In order to test the efficacy of our method, evaluations were done on two experimental scenarios with two datasets for each: action-only prediction and context-aware prediction. Our method achieved superior performance for predicting global activity classes and local action units.

1.1 Related Work

In general, there are three categories of works that are mostly related to ours: complex activity recognition, early detection of actions or events,[1] and event prediction in AI.

Complex Activity Recognition Recently, there has been a surge in interest in complex activity recognition by involving various structural information represented

---

[1]Concepts "action" and "event" are always interchangeably used in computer vision and other AI fields. In our discussion, we prefer to use "action" when referring human activity, and use "event" to refer more general things, such as "stock rising."

by spatial or temporal logical arrangements of several activity patterns. Most works aim to provide a good interpretation of complex activity. However, in many cases, inferring the goal of agents and predicting their plausible intended action are more desirable. Grammar based methods [24, 47] show effectiveness for composite human activity recognition. Pei et al. [42] proposed to deal with goal inference and intent prediction by parsing video events based on a stochastic context sensitive grammar (SGSG) which is automatically learned according to [48]. The construction of the hierarchical compositions of spatial and temporal relationships between the sub-events is the key contribution of their work. Without a formal differentiation between activity recognition and activity prediction, their system is actually doing an online detection of interesting events. Two important aspects for prediction, the earliness and the causality are missing in their discussion. The syntactic model is a very powerful tool for representing activities with high-level temporal logic complexity. Hamid et al. [20] proposed the idea that global structural information of human activities can be encoded using a subset of their local event sequences. They regarded discovering structure patterns of activity as a feature selection process. Although rich temporal structure information was encoded, they did not consider prediction possibility from that point.

Although not directly dealing with activity prediction, several notable works present various ways to handle activity structure. Logic based methods are powerful in incorporating human prior knowledge and have a simple inference mechanism [7]. To model temporal structure of decomposable activities, Gaidon et al. [17] and Niebles et al. [41] extended the classic bag-of-words model by including segmentation and dynamic matching. Kwak et al. [29] and Fan et al. [16] regarded complex activity recognition as a constrained optimization problem. Wang et al. [54] introduced the actionlet ensemble model, in which spatial structures of the features were encoded.

Early Detection of Action/Events It is important to distinguish between early detection and prediction. Essentially they are dealing with prediction in different semantic granularity. Early detection tries to recognize an ongoing atomic action from observation of its early stage. For example, an action of "handshaking" can be early detected by just observing "outstretched hand." However, for activity prediction, it tries to infer the intention or a higher level activity class with observation of only a few action units.

Ryoo [46] argued that the goal of activity prediction is to recognize unfinished single actions from observation of its early stage. Two extensions of bag-of-words paradigm, dynamic BoW and integral BoW are proposed to handle the sequential nature of human activities. Cao et al. [9] extended Ryoo's work to recognize human activities from partially observed videos, where an unobserved subsequence may occur at any time by yielding a temporal gap in the video. Hoai and De la Torre [22] proposed a max-margin early detector. They simulated the sequential frame-by-frame data arrival for training time series and learned an event detector that correctly classifies partially observed sequences. To deal with the sequential arrival of data, they developed a new monotonicity of detection function and formulated the problem within the structural support vector machine framework. Davis and Tyagi [12] addressed rapid recognition of human actions by the probability ratio test. This is a passive method for early detection. It assumes that a generative hidden Markov model for an event class, trained in a standard way, can also generate partial events.

Event Prediction in Other Fields While human activity prediction has received little attention in the computer vision field, predicting events or agent behaviors have been extensively studied in many other AI fields. Neill et al. [39] studied disease outbreak prediction. Their approach, like online change-point detection [13], is based on detecting the locations where abrupt statistical changes occur. Brown et al. [8] used the n-gram model for predictive typing, i.e., predicting the next word from previous words. Event prediction has also been studied in the context of spam filtering, where immediate and irreversible decisions must be made whenever an email arrives. Assuming spam messages were similar to one another, Haider et al. [19] developed a method for detecting batches of spam messages based on clustering. Financial forecasting [26] predicts the next day stock index based on the current and past observations. This technique cannot be directly used for activity prediction because it predicts the raw value of the next observation instead of recognizing the higher level event class. Kitani et al. [27] formulated the prediction task as a decision-making process [58] and proposed to forecast human behavior by leveraging the recent advances in semantic scene labeling [36, 37] and inverse optimal control [1, 4, 31, 57]. They predicted the destination of pedestrians and the routes they will choose.

2 Activity Prediction

To allow for more clarity in understanding our activity prediction model, we want to first provide an abstraction of activity prediction problem. Essentially, we transform the activity prediction problem into the problem of **early prediction on sequential data representation**. So the solution to this problem involves answering the following two questions: (1) "how to represent activity as a sequential data, which the way it is?"; (2) "how to do early prediction on such kind of representation?". We call the first one representation phase, and the second one prediction phase.

In the representation phase, an observation of complex human activity (e.g., from a camera, or from a rich networked sensor environments) is temporally segmented into semantic units in terms of component atomic actions (we call them actionlets). The boundaries between actionlets are detected by monitoring motion patterns (Sect. 3.1.1). Inside each segment, observed actionlets and objects are detected and quantized to symbolic labels which map to action and object classes.

In the prediction phase, the prediction problem becomes a sequence classification problem, but given only partial observation of the sequence (the beginning part). In data mining literature, sequence classification problem has three main categories of approaches: (1) **feature selection** with traditional vector based classification.

(e.g., K-grams as features); (2) **distance function**, which measures similarity of two sequences. (KNN or SVM kernel can be used as classifier in this scenario.) (3) **model based method**, which simulates a generative process to get a sequence. A model trained on sequences in one class can assign a likelihood to a new sequence. Specific models include K-order Markov model, variable order Markov model (VMM), and HMM.

Our approach proposes the prediction model in two scenarios: action-only and context-aware, which is characterized by what kind of information used for prediction. In both cases, we adopt the third strategy to train a model for prediction. The reasoning behind this choice is explained later. We formalize the representation first.

- Given an alphabet of actionlet symbols $\Sigma = \{a_1, a_2, a_3, \ldots, a_n\}$, an observation of activity is represented by a *simple symbolic sequence*, which is an ordered list of the actionlet symbols from the alphabet. An **action-only prediction model** takes this type of sequence of actionlets from an ongoing activity as input, and predicts the high-level activity class. For example, the activity "marriage proposal" is composed of four actionlets: $\langle a_{\text{hold-hands}}, b_{\text{kneel}}, c_{\text{kiss}}, d_{\text{put-ring-on}} \rangle$, shown in Fig. 7.2 (top).
- Given an alphabet of semantic symbols (actionlet and object labels) $\Sigma = \{e_1, e_2, e_3, \ldots, e_n\}$. In **context-aware prediction model**, an observation of activity is represented by a *complex symbolic sequence*, which is an ordered list of vectors. Each vector is a subset of the alphabet. For example, the activity "cook smashed potato dish" is composed of following itemsets: $\langle (abc)(cbd)(ed)f \rangle$, the meaning of each symbol is shown in Fig. 7.2 (bottom).

Now we specifically talk about our prediction models, and the reasons we chose them. For our action-only prediction model, we propose to use probabilistic suffix tree (PST; an implementation of VMM) as our main causality model. The reason for using this model is that the first two categories of approaches: feature-selection-based and distance-function-based, cannot handle partial sequence as input (i.e., useful patterns in the sequence that are highly dependent on the global observation). And among approaches in the third category, HMM can only model 1-order dependency, so it will ignore a lot of long-term causality information between activity components, which we believe is essential for prediction. K-order Markov model restricts order number to a specific order, so it will lack flexibility, making it unable to include small order and large order causalities at the same time, or in an adaptive fashion. So VMM is the most suitable model for early classification of sequence data, which models long-duration and short-duration sequential dependency as causality, and requires no need to see the whole sequence.

For our context-aware prediction model, we propose to use SPM to include objects cues for prediction. SPM is well suited for this problem because it uses item sets as sequence unit, so we can put co-occurrence of action and object as an enriched observation unit in complex activity scenario. Also, it can be easily tuned to fit into our whole prediction framework, which can be seen as an upgraded, rather compatible version of our action-only model. Details will be discussed in the later sections. Table 7.1 summarizes the capability of different methods.

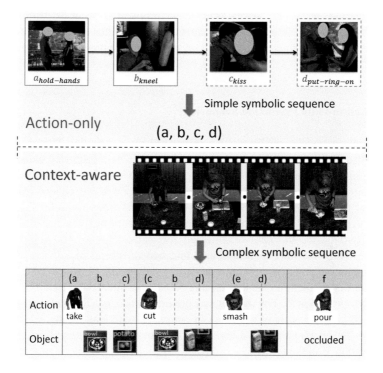

Fig. 7.2 Two scenarios of activity modeling: (*Top*) action-only model and (*Bottom*) context-aware model

Table 7.1 Comparison of different methods in terms of capability for modeling activity prediction

Models	Sequence classification	Partial sequence classification (prediction)	Causality	Context
Feature based	+	−	−	−
Distance function based	+	−	−	−
HMM	+	+	+ (of order 1)	−
Variable order Markov model	+	+	+	−
Sequential pattern mining	+	+	+	+

2.1 Two Prediction Scenarios

To demonstrate the effectiveness of our proposed approach, we evaluate prediction tasks on two datasets for two different scenarios. For the action-only prediction, we test the ability of our approach to predict human daily activities, such as "making a phone call," which have middle-level temporal complexity. Next, we test our model at high-level temporal complexity activities on a tennis game dataset collected by the authors. For the context-aware prediction, first we test our approach on a cooking activity dataset, where the whole kitchen environment is observed from a static camera. Second, we extend our data domain to a networked sensor environments, where we want to predict several kinds of morning activities in an office lounge, such as "coffee time."

2.1.1 Action-Only Prediction Scenario

Our prediction model is applicable to a variety of human activities. The key requirement is that the activity should have multiple steps where each step constitutes a meaningful action unit. Without loss of generality, we choose two datasets with significant different temporal structure complexity. First, we collect real world videos for tennis games between two top male players from YouTube. Each point with an exchange of several strokes is considered as an activity instance, which involves two agents. In total, we collected 160 video clips for 160 points from a 4-h game. The clips were then separated into two categories of activity, where 80 clips are winning points and 80 clips are losing points with respect to each specific player. So our prediction problem on this dataset becomes the question: "Can we predict who will win?". The dataset and prediction task are illustrated in Fig. 7.3. Since each point consists of sequence of actionlets with length ranging from 1 to more than 20, tennis game has a high-level temporal structure complexity in terms of both variance and order.

Second, we choose the Maryland human-object interactions (MHOI) dataset [18], which consists of six annotated activities: *answering a phone call, making a phone call, drinking water, lighting a flash, pouring water into container,* and *spraying*. These activities have about 3–5 action units each. Constituent action units share similar human movements: (1) reaching for an object of interest, (2) grasping the object, (3) manipulating the object, and (4) put back the object. For each activity, we have 8–10 video samples. There are 54 video clips in total. Examples in this dataset are shown in Fig. 7.5.

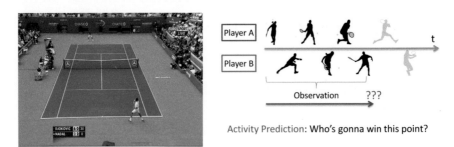

Fig. 7.3 *Left*: tennis game dataset. *Right*: activity prediction task on this dataset

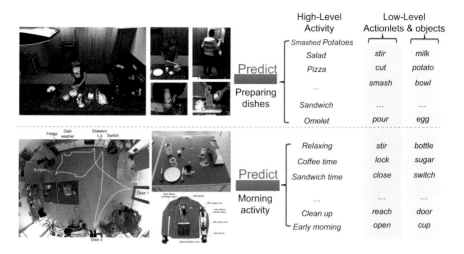

Fig. 7.4 Two datasets for context-aware prediction scenario. *Top*: MPII-Cooking dataset [44], where we want to predict the type of dish the subject is preparing. *Bottom*: UCI-OPPORTUNITY dataset [43], where we want to predict current ongoing morning activity

2.1.2   Context-Aware Prediction Scenario

To verify our context-aware model, we perform experiments on two complex activity datasets, where human actions involve a lot of interactions with various objects. The first is a fine-grained cooking activity dataset, and the other is a complex morning activity dataset in highly rich networked sensor environment.

The MPII Cooking Activities Dataset (MPII-Cooking) [44] contains 44 instances of cooking activity, which are continuously recorded in a realistic setting, as shown in Fig. 7.4 (top). Predictable high-level activities are about preparing 14 kinds of dishes, including: *making a sandwich*, *making a pizza*, and *making an omelet*, etc. There are overall 65 different actionlets as building blocks shared among various cooking activities, such as *cut*, *pour*, *shake*, and *peel*.

The OPPORTUNITY Activity Recognition Dataset (UCI-OPPORTUNITY) [43] was created in a sensor-rich environment for the machine recognition of human activities, as shown in Fig. 7.4 (bottom). They deployed 72 sensors of 10 modalities in 15 wireless and wired networked sensor systems in the environment, on the objects, and on the human body. The data are acquired from 12 subjects performing morning activities, yielding over 25 h of sensor data. It contains 5 high-level predictable activities (*Relaxing, Coffee time, Early Morning, Cleanup, Sandwich time*), 13 low level actionlets (e.g., *lock, stir, open, release*), and 23 interactive objects (e.g., *bread, table, glass*).

3 Proposed Approach

3.1 Representation: Activity Encoding

3.1.1 Actionlets Detection by Motion Velocity

Temporal decomposition is the first key step for our representation of complex activity. It is used to find the frame indices that can segment a long sequence of human activity video into multiple meaningful atomic actions. Relevant work can be found in [51]. We call these atomic actions **actionlets**. We found that the velocity changes of human actions have similar periodic regularity. Figure 7.5 shows three examples of actionlet segmentation and detection.

The specific method includes these steps: (1) use Harris corner detector to find significant key points; (2) use Lucas-Kanade (LK) optical flow to generate the trajectories for key points; (3) for each frame, accumulate the trajectories/tracks at these points to get a velocity magnitude:

$$V_t = \sum_{p(x_{i,t}, y_{i,t}) \in F_t} \sqrt{(x_{i,t} - x_{i,t-1})^2 + (y_{i,t} - y_{i,t-1})^2}, \qquad (7.1)$$

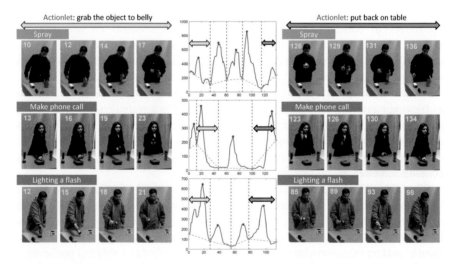

Fig. 7.5 Actionlet detection by motion velocity. Time series figures show motion velocity changes over time (smoothed). The *horizontal axis* is the frame index, and the *vertical axis* is the velocity strength computed according to formula (7.1) *Red dots* on each local peak indicate actionlets centers, and *green dots* on each local valley indicate the segmentation points. A point is considered an actionlet center only if it has the local maximal value, and was preceded (to the *left*) by a value lower by a threshold. Actionlets are obtained by extracting segments between two consecutive *vertical dashed lines*. Two actionlets shared by three different types of actions are shown as examples: "*grab the object to belly*" (*left*) and "*put back on the table*" (*right*). Images on *each row* are from the same video marked with frame indices

where V_t represents the overall motion velocity at frame F_t, p_i is the ith interest point found in frame F_t. $(x_{i,t}, y_{i,t})$ is the position of point p_i in the frame. We observed that each hill in the graph represents a meaningful atomic action. For each atomic action, the start frame and the end frame always have the lowest movement velocity. The velocity reaches the peak at the intermediate stage of each actionlet. To evaluate our temporal decomposition approach, a target window with the size of 15 frames around the human labeled segmentation point is used as the ground truth. We manually labeled 137 segmentation points for all 54 videos in the MHOI dataset. The accuracy of automatic actionlets segmentation is 0.83. For the tennis game dataset, we cut the video clips into top-half and bottom-half to handle actionlets of two players. We labeled 40 videos with 253 actionlets in it. The actionlet segmentation accuracy is 0.82.

3.1.2 Activity Encoding

Based on accurate temporal decomposition results, we can easily cluster actionlet into meaningful groups so that each activity can be represented by a sequence of actionlets in a syntactic way. A variety of video descriptors can be used here as long as they can provide discriminative representations for the actionlets.

Due to different spatial extent of humans in the scene and different background motion styles, two approaches are used to compute descriptors for tennis game dataset and MHOI dataset, respectively. For the MHOI dataset which has a large scale human in the scene and a static background, we use the 3-D Harris corner detector to find sparse interest points. Each local area is described by HoG (Histogram of Gradients) and HoF (Histogram of Flow) descriptors [30]. Furthermore, we vector quantize the descriptors by computing memberships with respect to a descriptor codebook of size 500, which is obtained by k-means clustering of the descriptors. Then, actionlets categories are learned from histogram of spatial-temporal words using an unsupervised algorithm [40]. To evaluate the actionlets encoding results, human experts watch video segments corresponding to each actionlet, and annotate them according to their semantic meanings, such as "reach the object," "grab the object to belly," and "grab the object to head," etc. The Rand index[2] of clustering is 0.81.

For the tennis game dataset, the scale of player in the video is very small, therefore it is difficult to get sufficient local features by using sparse sampling methods. Here, we use dense trajectories [53] to encode actionlets. For every actionlet, we sample the initial feature points every w pixels at multiple spatial scales. All tracking points are obtained by a median filter in a dense optical flow field from the points in the previous frame. For each trajectory, the descriptor is

---

[2]Rand index is a measure of the similarity between data clustering and ground truth. It has a value between 0 and 1, with 0 indicating that the two data clusters do not agree on any pair of points and 1 indicating that the data clusters are exactly the same.

calculated in a 3-D volume. Each such volume is divided into sub-volumes. HOG, HOF, and MBH features are then computed for every sub-volume. In our approach, we use the same parameters indicated in [53]. The codebook size we used is 1000. In addition, to remove the noises caused by camera movements and shadows, a human tracker [11] is used before extracting feature records. For evaluation, we group 253 actionlets from 40 annotated videos into 10 categories, and the Rand index of clustering is 0.73. For MPII-Cooking and UCI-OPPORTUNITY datasets, we use pre-annotated actionlet boundaries provided by the dataset when we perform activity encoding. Our focus on these two datasets is to evaluate context-aware prediction model.

3.2 Action-Only Causality Model

Here we introduce the model of human activity prediction, which is illustrated in Fig. 7.1. Let Σ be the finite set of actionlets, which are learned from videos using unsupervised segmentation and clustering methods. Let $D_{\text{training}} = \{r^1, r^2, \ldots, r^m\}$ be the training sample set of m sequences over the actionlet alphabet Σ, where the length of the ith ($i = 1, \ldots, m$) sequence is l_i (i.e., $r^i = r^i_1 r^i_2 \ldots r^i_{l_i}$, where $r^i_j \in \Sigma$). Based on D_{training}, the goal is to learn a model P that provides a probability assignment $p(t)$ for an ongoing actionlet sequence $t = t_1, t_2, \ldots, t_{\|t\|}$. To realize this design with maximum predictive power, we include two sources of information in the model. One is the causality cue hidden in the actionlet sequences, which encodes the knowledge about the activity. The other is the unique predictable characteristic for each kind of human activity, which answers the questions why a particular activity can be predicted and how early an activity can be predicted with satisfactory accuracy.

Causality is an important cue for human activity prediction. Our goal is to automatically acquire the causality relationships from sequential actionlets. VMM [5] is a category of algorithms for prediction of discrete sequences. It suits the activity prediction problem well, because it can capture both large and small order Markov dependencies extracted from training data. Therefore, it can encode richer and more flexible causal relationships. Here, we model complex human activity as a PST [45] which implements the single best L-bounded VMM (VMMs of degree L or less) in a fast and efficient way.

The goal of the PST learning algorithm is to generate a conditional probability distribution $\gamma_s(\sigma)$ to associate a "meaningful" context $s \in \Sigma^\star$ with the next possible actionlet $\sigma \in \Sigma$. We call the function $\gamma_s(\sigma)$ the *next symbol probability function*, and denote the trained PST model as $\overline{T}$, with corresponding suffix set as $\overline{S}$ consisting of actionlets sequence of all the nodes. Algorithm 1 shows the detailed building process of PST, where there are five user specified parameters. Figure 7.6 shows an example PST constructed from a training sequence of actionlets.

Algorithm 1 Construction of L-Bounded PST $\overline{T}$ $(L, P_{\min}, \alpha, \beta, \lambda)$

---

1. **Forming candidate suffix set $\overline{S}$**: Let $D_{\text{training}} = \{r^1, r^2, \ldots, r^m\}$ be the training set, and assume s is a subsequence of r^i $(i = 1, \ldots, m)$. If $|s| < L$ and $P(s) > P_{\min}$, then put s in $\overline{S}$. $P_{\min}$ is a user specified minimal probability requirement for an eligible candidate. $P(s)$ is computed from frequency count.
2. **Testing every candidate $s \in \overline{S}$**: For any $s \in \overline{S}$, test following two conditions:

   - (1) $P(\sigma|s) \geq \alpha$, which means the context subsequence s is meaningful for some actionlet σ. Here, α is defined by user to threshold a conditional appearance.
   - (2) $\frac{P(\sigma|s)}{P(\sigma|\text{suf}(s))} \geq \beta$, or $\leq 1/\beta$, which means the context s provides extra information in predicting σ relative to its longest suffix $\text{suf}(s)$. β is a user specified threshold to measure the difference between the candidate and its direct parent node.
   - **Then**, if s passes above two tests, add s and its suffixes into $\overline{T}$.

3. **Smoothing the probability distributions to obtain $\gamma_s(\sigma)$**:
   For each s labeling a node in $\overline{T}$, if $P(\sigma|s) = 0$, we assign a minimum probability λ. In general, the *next symbol probability function* can be written as:
   $\gamma_s(\sigma) = (1 - |\Sigma|\lambda)P(\sigma|s) + \lambda$. Here, λ is the smoothing factor defined empirically.

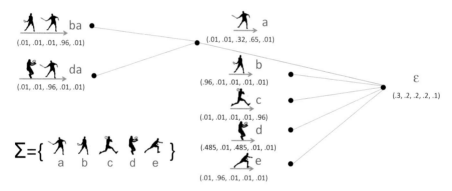

Fig. 7.6 An example PST corresponding to the training sequence $r = badacebadc$ over alphabet $\Sigma = \{a, b, c, d, e\}$. The vector under each node is the probability distribution over alphabet associated with the actionlets subsequence (in *red*) (e.g., the probability to observe d after a subsequence, whose largest suffix in the tree is ba, is 0.96)

3.3   Context-Aware Causality Model

The context-aware causality model embodies vocabularies of visual elements including actionlets and objects as enriched symbolic sequences that specify syntactic (compositional) relationships. We call each unit of the enriched symbolic sequence eventlet.[3] SPM [34, 56] was first introduced in the area of data mining,

---

[3]In this chapter, we use eventlet to refer to observation of actionlet and objects co-occurrence. An eventlet $e = \langle \{a^*\} \bigcup \{o_1, o_2, \ldots, o_m\} \rangle$, where a^* represents a particular actionlet, and o_i represents a particular object interacting with a^* within its segment. In our case, n will always

where a sequence database stores a set of records consisting of sequences of ordered events, with or without concrete notions of time. SPM algorithm mines the sequence database looking for repeating patterns (known as frequent sequences) that can be used later by the end-users to find associations between different items or events in their data for purposes such as marketing campaigns, web usage mining [15, 38, 50], DNA sequences analysis [52], prediction and planning. There are two main reasons we propose to use SPM as our context-aware model. First, our framework detects a sequence of visual concepts, which are well quantized as semantical labels with concrete time stamps. Essentially the data structure we are dealing with is quite similar to the common data types in the sequential database, such as customer transactions data or purchase sequences in a grocery store. Second, in our context-aware scenario, actionlets have been enriched with co-occurred objects, so the newly formed eventlet sequence has both the compositional and sequential nature, which can be perfectly fitted into an SPM model.

We present two algorithms, which we call *Mapping-based* (Algorithm 2) and *SPM-based* (Algorithm 3). The *mapping-based* algorithm is a simplified version of our context-aware causality model, which transforms the complex symbolic sequence (eventlets) into a simple symbolic sequence. Then a similar PST model can be applied for the mapped simple symbolic sequences, as we used in the action-only model.

The *SPM-based* algorithm is a relatively complex version of our context-aware causality model, which finds frequent subsequence of itemsets as sequential patterns. Then we utilize the mined sequential patterns to compute the conditional probabilities $\gamma_s(\sigma)$, which associates a "meaningful" context $s \in \Sigma^*$ with next possible eventlet. In our context-aware model as shown in Fig. 7.1, we have the following definitions.

Algorithm 2 Mapping-Based Context-Aware Model
___

1. **Frequent Itemsets Phase**:
   Given ξ_{itemset}, find the set of all frequent itemsets *FI* applying Apriori algorithms [3].
   Notice: A small difference here is that the support of an itemset is defined as the fraction of activity instances (the sequences of eventlets) rather the fraction of eventlets.
2. **Mapping Phase**:
   $f : FI \rightarrow FI'$, where $FI' \subset Z$, frequent itemset $i_s \in FI$ is mapped into an integer $x \in FI'$.
3. **Transformation Phase**: FI' is further broken down into individual interactions of frequent itemsets along the time line, e.g. the examples in Table 7.2.
   Notice: The reason for this mapping is that by treating frequent itemsets as single entities, we can transform our context-aware observation, a complex symbolic sequence, into a simple symbolic sequence representation. This transformed representation is called D_{training}.
4. **Construct Causality Model**: Use the set D_{training} to build the causality model by calling Algorithm 1.

---

be 0, 1, or 2 with the meaning of none, one, or two co-occurrent interacting objects (we assume one person at most can operate two different objects at the same time with two hands).

Algorithm 3 SPM-Based Context-Aware Model

1. **Frequent Itemsets Phase**:
   Given ξ_{itemset}, find the set of all frequent itemsets FI applying Apriori algorithms [3].
2. **Mapping Phase**:
   $f : FI \rightarrow FI'$, where $FI' \subset Z$, frequent itemset $i_s \in FI$ is mapped into an integer $x \in FI'$.
3. **SPM Phase**:
   Initialize $SP_1 = FI'$. Use algorithm AprioriAll[2] to find $SP_2, \ldots, SP_k$, where k is the largest length of frequent sequential pattern.
4. **Construct Causality Model**: Based on mined $SP_1, \ldots, SP_k$ and corresponding support for each sequential pattern in it, causality model $P(\sigma|s)$ can be computed by Bayes rules $P(\sigma|s)$, assume $|s| = k'$.

   - If $s\sigma \in SP_{k'+1}$, $P(s\sigma) = \frac{\sup_D(s\sigma)}{\sum_{x_i \in SP_{k'+1}} \sup_D(x_i)}$, $P(s) = \frac{\sup_D(s)}{\sum_{x_i \in SP_{k'}} \sup_D(x_i)}$, $P(\sigma|s) = \frac{P(s\sigma)}{P(s)}$.
   - Otherwise, $P(\sigma|s) = 0$.

5. **Smoothing the probability distributions to obtain „ₛ(ff)**:
   For each s, if $P(\sigma|s) = 0$, we assign a minimum probability λ. In general, the *next symbol probability function* can be written as:
   $\gamma_s(\sigma) = (1 - |\Sigma|\lambda)P(\sigma|s) + \lambda$. Here, λ is the smoothing factor defined empirically.

Table 7.2 Transformed representation: mapping complex symbolic sequence into simple symbolic sequence

Activity instance ID	Original sequence	Frequent itemset and mapping	After transformation
1	$\langle a, b \rangle$	$a \rightarrow 1, b \rightarrow 5$	$\langle 1, 5 \rangle$
2	$\langle (cd), a, (efg) \rangle$	$a \rightarrow 1, e \rightarrow 2, g \rightarrow 3, (eg) \rightarrow 4$	$\langle 1, 2 \rangle, \langle 1, 3 \rangle, \langle 1, 4 \rangle$
3	$\langle (ahg) \rangle$	$a \rightarrow 1, g \rightarrow 3$	$\langle 1 \rangle, \langle 3 \rangle$
4	$\langle a, (eg), b \rangle$	$a \rightarrow 1, e \rightarrow 2, g \rightarrow 3, (eg) \rightarrow 4, b \rightarrow 5$	$\langle 1, 2, 5 \rangle, \langle 1, 3, 5 \rangle, \langle 1, 4, 5 \rangle$
5	$\langle b \rangle$	$b \rightarrow 5$	$\langle 5 \rangle$

- A set of observations of human activity, saying activity archive D, is represented as a set of sequential records of eventlets. Each eventlet is segmented according to actionlet boundaries, and represented as an itemset of detected visual concepts.
- An eventlet sequence is an ordered list of itemsets, for example, $s = \langle a(be)c(ad) \rangle$. An itemset is a set drawn from items in Σ, and denoted $(i_1, i_2, \ldots, i_k)$, such as a and (be) in the previous example. Σ is a set of N unique items $\Sigma = i_1, i_2, \ldots, i_N$, where each element in Σ can be either an actionlet label or an object label.
- The support of itemset or eventlet, $i_s = (i_1, i_2, \ldots, i_k)$ is defined as the fraction of activity instances (the sequences of eventlets) $s \in D$ that contains the itemset in any one of its possibly many eventlets. Given a support threshold *min_sup* ξ_{itemset}, an itemset is called frequent itemset on D if $\sup_D(i_s) \geq \xi_{\text{itemset}}$. The set of all frequent itemsets is denoted as FI.
- The support of a sequence s_a is defined as the fraction of activity instances $s \in D$ that contains s_a, denoted by $\sup_D(s_a)$. Given a support threshold

min_sup ξ_{sequence}, a sequence is called a frequent sequential pattern on D if $\text{sup}_D(s_a) \geq \xi_{\text{sequence}}$.
- The length of a sequence is the number of itemsets in the sequence. SP_k denotes the set of frequent sequential pattern with length k. So $SP_1 = FI$.
- Either based on Algorithm 2 or Algorithm 3, we can now train a context-aware causality model $\gamma_s(\sigma)$, as we did in the action-only case.

3.4 Predictive Accumulative Function

In this section, we want to answer "why a particular activity can be predicted," and how to make our model automatically adapted to activities with different predictability. For example, "tennis game" is a late-predictable problem in the sense that a long sequence of actionlets performed by two players was observed, the last several strokes will strongly impact the winning or losing results. In contrast, "drinking water" is an early predictable problem, since as long as we observed the first actionlet "grabbing a cup," we probably can guess the intention. To characterize the predictability of activities, we formulate a PAF. Different activities usually reflect very different PAFs. In our model, PAF can be learned automatically from the training data. For activity prediction at a later stage, we use PAF to weight the observed patterns in every stage of ongoing sequence.

Suppose $k \in [0, 1]$ indicates the fraction of beginning portion (prefix) of any sequence. D is the training set. Let D_k be the set of sequences, where each sequence consists of the first k percentage of the corresponding $r = (r_1, r_2, \ldots, r_l) \in D$, where $r_i(i = 1, 2, \ldots, l) \in \Sigma$, l is the length of r. We use $r_{\text{pre}(k)}$ to represent the corresponding "prefix" sequence of r in D_k. Obviously $|D| = |D_k|$.

Given the first k percentage of the sequence observed, the information we gain can be defined as follows:

$$y_k = \frac{H(D) - H(D|D_k)}{H(D)}. \tag{7.2}$$

Here the entropy $H(D)$ evaluates the uncertainty of a whole sequence, when no element is observed, and the conditional entropy $H(D|D_k)$ evaluates the remaining uncertainty of a sequence after first k percentage of sequence is checked.

$$H(D) = -\sum_{r \in D} p^T(r) \log p^T(r),$$

$$H(D|D_k) = -\sum_{r_{\text{pre}(k)} \in D_k} \sum_{r \in D} p^T(r, r_{\text{pre}(k)}) \log p^T(r|r_{\text{pre}(k)}). \tag{7.3}$$

Since $r_{\mathrm{pre}(k)}$ is the "prefix" of r, we have

$$p^{\overline{T}}(r, r_{\mathrm{pre}(k)}) = p^{\overline{T}}(r), \text{ and}$$

$$p^{\overline{T}}(r|r_{\mathrm{pre}(k)}) = \frac{p^{\overline{T}}(r, r_{\mathrm{pre}(k)})}{p^{\overline{T}}(r_{\mathrm{pre}(k)})} = \frac{p^{\overline{T}}(r)}{p^{\overline{T}}(r_{\mathrm{pre}(k)})}. \tag{7.4}$$

From trained PST model $\overline{T}$, we write

$$p^{\overline{T}}(r) = \prod_{j=1}^{\|r\|} \gamma_{s^{j-1}}(r_j), \text{ and}$$

$$p^{\overline{T}}(r_{\mathrm{pre}(k)}) = \prod_{j=1}^{\|r_{\mathrm{pre}(k)}\|} \gamma_{s^{j-1}}(r_{\mathrm{pre}(k)_j}). \tag{7.5}$$

The nodes of $\overline{T}$ are labeled by pairs (s, γ_s), where s is the string associated with the walk starting from that node and ending in tree root; and $\gamma_s : \Sigma \to [0, 1]$ is the *next symbol probability function* related with s, $\sum_{\sigma \in \Sigma} \gamma_s(\sigma) = 1$.

Based on above discussions, we can have a sequence of data pair (k, y_k) by sampling $k \in [0, 1]$ evenly from 0 to 1. For example, by using 5% as interval, we will collect 20 data pairs. Now we can fit a function f_p between variable k and y, which we call PAF: $\mathbf{y} = \mathbf{f_p(k)}$. Function f_p depicts the predictable characteristic of a particular activity. Figure 7.7 shows PAFs in two extreme cases. The curves are generated on simulated data to represent an early predictable problem and a late predictable problem, respectively.

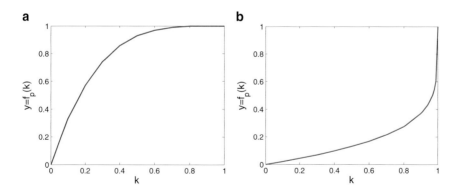

Fig. 7.7 PAFs for depicting predictable characteristics of different activities. (**a**) Early predictable problem. (**b**) Late predictable problem

3.5   Final Prediction Model

Given an ongoing sequence $t = t_1, t_2, \ldots, t_{\|t\|}$, we can now construct our prediction function by using the knowledge learned from Sects. 3.2 and 3.3 (causality) and Sect. 3.4 (predictability):

$$p^{\overline{T}}(t) = \sum_{j=1}^{\|t\|} f_p \left(\frac{\|t_1 t_2 \ldots t_j\|}{\|t\|} \right) \log \gamma_{s^{j-1}}(t_j), \tag{7.6}$$

which computes the weighted log-likelihood of t as the prediction score with the knowledge of trained PST model $\overline{T}$ and learned PAF f_p.

Given an observed ongoing sequence of actionlets, our ultimate goal is to predict the activity class it belongs to. This problem can fit into the context of supervised classification where each class $c(c = 1, \ldots, C)$ is associated with a prediction model $p^{\overline{T}_c}(t)$ for which the empirical probabilities are computed over the whole set of sequences of this class belonging to the training set. Given an ongoing sequence $t = t_1, t_2, \ldots, t_{\|t\|}$, the sequence t is assigned to the class $c0$ corresponding to the prediction model $p^{\overline{T}_{c0}}$ for which maximal prediction score has been obtained: $\mathbf{p^{\overline{T}_{c0}}(t) = Max\{p^{\overline{T}_c}(t), c = 1, \ldots, C\}}$.

4   Experiments

We present experimental results on two scenarios of activity prediction: action-only and context-aware.

4.1   Action-Only Activity Prediction

4.1.1   Middle-Level Complex Activity Prediction

Samples in MHOI dataset are about daily activities (e.g., "making phone call"). This type of activity usually consists of 3–5 actionlets and lasts about 5–8 s, so we call it middle-level complex activity. In this dataset, each category has 8–10 samples. We evaluate the prediction accuracy by using the standard "leave-one-out" method, and fit activity prediction in the context of multi-class classification problem. Different from traditional classification task, for activity prediction, we focus on the predictive power of each method. The goal is to use an observation ratio as small as possible to make an accurate prediction. To train a prediction model, we constructed an order 5-bounded PST and fit a PAF, respectively. We compare our method of activity prediction with existing alternatives, including: (1) Dynamic Bag-of-Words model

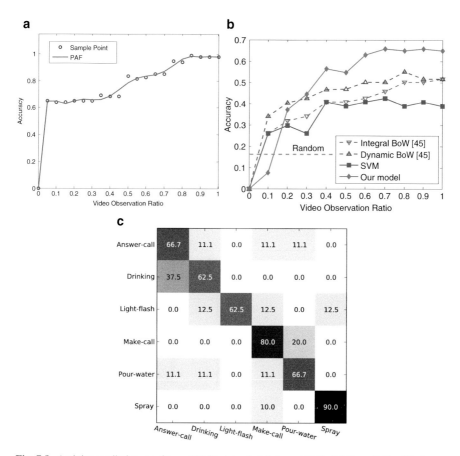

Fig. 7.8 Activity prediction results on MHOI dataset. (**a**) shows PAF of daily activity. (**b**) shows comparison of prediction accuracy of different methods. A higher graph suggests that the corresponding method is able to recognize activities more accurately and earlier than the ones below it. Our approach shows the best performance. (**c**) shows the confusion matrix at 60 % of observation

[46], (2) Integral Bag-of-Words model [46], and (3) a basic SVM-based approach. All baseline methods are adopting the same bag-of-words representation with a codebook size of 500, which is built from the local features HOG/HOF.

Figure 7.8a illustrates the process of fitting PAF from training data. It shows that daily activities such as examples from MHOI dataset are early predictable. That means the semantic information at early stage strongly exposes the intension of the whole activity. Figure 7.8b illustrates the performance curves of the implemented four methods. The results are averaged over six activities. Its horizontal axis corresponds to the observed ratio of the testing videos, while the vertical axis corresponds to the activity recognition accuracy. The figure confirms that the proposed method has great advantages over other methods. For example, after half of the video is observed (about 2 actionlets), our model is able to make a prediction with the accuracy of 0.6.

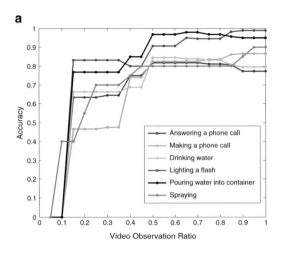

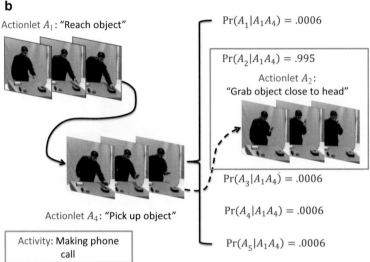

Fig. 7.9 Global and local prediction for a particular activity in MHOI dataset. (**a**) Prediction on six activities. (**b**) Predict next actionlet

Figure 7.9a shows detailed performance of our approach over 6 different daily activities in a binary classification setting. From the figure, we can see that the activity "Pouring water into container" has the best prediction accuracy and earliness. In this dataset, after the actors reach the object, they usually grab the object and put it close to the head. Three activities ("making a phone call," "drinking water," and "answering a phone call") share this process in the initial phase of the activity. So, in the activity "pouring water into container," after the first common actionlet "reach object," the second and third constituent actionlets make the sequence pattern quite distinctive. Besides predicting global activity classes, our model can also make local predictions. That means given observed actionlet sequence as context, the model can predict the most probable next actionlet. Figure 7.9b shows an example from our experiment results.

4.1.2 High-Level Complex Activity Prediction

In this experiment, we aim to test the ability of our model to leverage the temporal structure of human activity. Each sample video in the tennis game dataset is corresponding to a point which consists of a sequence of actionlets (strokes). The length of actionlet sequence of each point can vary from 1 to more than 20. So the duration of some sample videos may be as long as 30 s. We group samples into two categories, winning and losing, with respect to a specific player. Overall, we have 80 positive and 80 negative samples, respectively. Then a 6-bounded PST and a PAF are trained from data to construct the prediction model. The same "leave-one-out" method is used for evaluation.

Figure 7.10a illustrates the fitted PAF for tennis activity. It shows that tennis games are late predictable. That means the semantic information at late stage strongly impacts the results of classification. This is consistent with common sense about tennis games. Figure 7.10b shows prediction performance of our method. Here we compare two versions of our model to illustrate the improvement caused by considering predictable characteristic of activity. Since all other three methods, D-BoW, I-BoW and SVM, failed in prediction on this dataset, we did not show comparisons. In short, our model is the only one that has the capability to predict on high-level complex activity. Table 7.3 shows detailed comparisons of 7 methods on two datasets, where we include 3 sequence classification methods with each one representing a category of approaches. The details about other three methods will be discussed in Sect. 3.2.

4.1.3 Model Parameters

Advantage of our approach is that there are very few model parameters need to be tuned. Among them, the order L of PST is the most important one, since it determines the order of causal relationships that we want to incorporate. Figure 7.11 shows the impact of parameter selections on prediction accuracy. We can see higher order of PST performs better. This is because it includes long-term and short-term Markov dependencies at the same time.

Fig. 7.10 Activity prediction results on tennis game dataset. (**a**) shows PAF of tennis game. (**b**) shows prediction performance of our model. We did not show comparisons with the other three methods because of their inability to handle high-level complex activity, such as tennis game. Their prediction curves are nearly random, since Bag-of-Words representation is not discriminative anymore in this situation

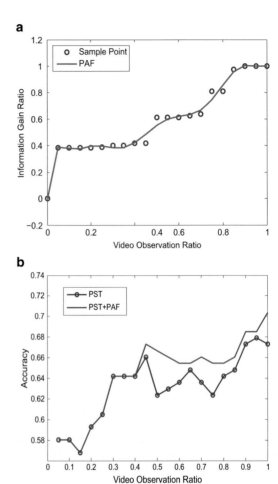

4.2 Context-Aware Activity Prediction

In this section, we report experimental results on context-aware situations, which demonstrate that presence of interactive objects may significantly improve activity prediction performance by providing discriminative contextual cues often at very early stage of the activity progress.

4.2.1 Experiments on Cooking Activity Dataset

Samples in MPII-Cooking dataset are about preparing various dishes (e.g., *making a sandwich*). Because of the varying degrees of complexity of different dishes, the length of eventlets sequence of each sample can vary from 20 to more than 150.

Table 7.3 Performance comparisons on two datasets

Methods	Tennis game dataset					MHOI dataset				
	20 %	40 %	60 %	80 %	100 %	20 %	40 %	60 %	80 %	100 %
Integral BoW [46]	0.52	0.49	0.47	0.48	0.44	0.32	0.41	0.42	0.50	0.52
Dynamic BoW [46]	0.53	0.51	0.47	0.49	0.53	**0.40**	0.47	0.50	0.55	0.52
BoW+SVM	0.56	0.52	0.51	0.48	0.49	0.30	0.41	0.41	0.39	0.39
Feature-based model (K-grams) [14]	0.45	0.54	0.51	0.47	0.43	0.33	0.38	0.40	0.47	0.49
Distance function base model [49]	0.52	0.56	0.48	0.49	0.48	0.14	0.27	0.51	0.53	0.53
HMM	0.43	0.46	0.51	0.46	0.58	0.23	0.38	0.56	0.47	0.43
Our action-only model	**0.59**	**0.64**	**0.65**	**0.65**	**0.70**	0.37	**0.57**	**0.63**	**0.65**	**0.65**

Random guess for MHOI dataset and tennis game dataset are 0.167 and 0.5, respectively. Actually comparison methods perform random guess on tennis game (Percentage as observation ratios). The bold values indicate the best result (i.e., highest accuracy) in each setting.

The average sequence length is 67.[4] For a particular activity, similar to experimental settings in action-only situation, we use all the samples in that category as the set of positive examples, and randomly select equal number of samples from remaining categories as the set of negative examples. Then we use the prediction task in the context of supervised binary classification of sequence data with varying observation ratios. To train an mapping-based context-aware model, we set $\xi_{itemset} = 0.3$, and put the transformed representation into a 11-order bounded PST to obtain a causality model. The PAF is also generated from the transformed representation according to the same method as before. To train a SPM-based context-aware model, we set $\xi_{itemset} = 0.3$ and $\xi_{sequence} = 0.2$.

To show the advantages of the proposed method, we compare our results with other three sequence classification methods, which may represent each of the three main categories of approaches. The details of comparison methods are as follows.

- ***k*-gram** ($k = 2$) with linear-SVM [14] represents feature-base methods. k-gram is the most popular feature selection approach in symbolic sequence classification.
- The **Smith-Waterman algorithm** with KNN [49] represents sequence distance-based methods. Since in our prediction task, for most cases, we want to utilize partial sequence as observation. Then it always needs to compute distances between a long sequence (from training set in a lazy learning style) and a short sequence (from the beginning part of the sample to be predicted), so local alignment based distance [49] is preferred.
- Discrete **Hidden Markov Model** (D-HMM) represents generative model based methods.

---

[4]Notice that there are many situations that some periodical actions will be segmented to consecutive duplicate eventlets, e.g. action "cut."

Fig. 7.11 Comparison of
different order
bounded-PSTs. *Top*:
prediction performance on
tennis game dataset. *Bottom*:
prediction performance on
MHOI dataset

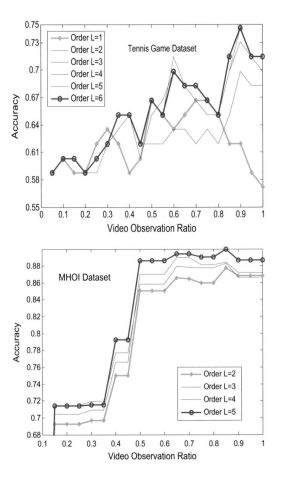

Figure 7.12a and Table 7.4 show comparison of prediction accuracy of different
methods. We can clearly see that the context information, such as interactive objects,
can perform as a strong cue for predicting activity type.

In HMM, high order dependencies between actionlets are lost. Though it still can
get satisfactory classification performance at full observation, in terms of prediction,
it has obvious limitations. In feature-based and distance-based approaches, the
sequential nature of human activity cannot be captured. The context information
cannot be added either. Therefore, without causality (sequential patterns) and
context (object cues), things become unpredictable. Because in MPII-Cooking
dataset, many actionlets, such as *cut* and *take*, are actually routines in preparing
different dishes, ignoring sequential patterns may result in confusion between
activity classes.

Fig. 7.12 Performance comparisons on context-aware prediction evaluation. (**a**) Activity prediction on MPII-Cooking dataset. (**b**) Activity prediction on UCI-OPPORTUNITY dataset

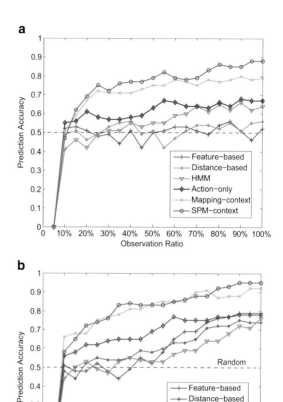

4.2.2 Experiments on Sensor Dataset

In this experiment, we aim to test the ability of our model to leverage from visual domain to sensor networking environment. Since this dataset has no pre-segmented activity video clips, we first locate samples of each kind of high-level activity based on their annotations, such as starting and ending time stamps. Then we extract actionlets and objects labels from sensor data within the time interval of each sample. For mapping-based context-aware model, we set $\xi_{itemset} = 0.2$ and maximum PST order as 7. For SPM-based context-aware model, we set $\xi_{itemset} = 0.2$ and $\xi_{sequence} = 0.2$. We also compare our approaches with other three methods mentioned above.

Figure 7.12b and Table 7.4 show comparison of prediction accuracy of different methods. From the figure, we can see that the sensor dataset is relatively "easier"

Table 7.4 Performance comparisons on two context-aware datasets

Methods	Cooking activity dataset					Sensor dataset				
	20 %	40 %	60 %	80 %	100 %	20 %	40 %	60 %	80 %	100 %
Feature-based model (K-grams) [14]	0.48	0.42	0.53	0.56	0.55	0.52	0.55	0.69	0.77	0.79
Distance function base model [49]	0.5	0.49	0.51	0.55	0.54	0.55	0.59	0.63	0.72	0.76
HMM	0.49	0.53	0.60	0.62	0.65	0.49	0.54	0.59	0.68	0.77
Action-only model	0.58	0.59	0.64	0.64	0.66	0.62	0.65	0.75	0.77	0.78
Context-aware model (Mapping)	**0.67**	**0.71**	**0.78**	**0.77**	**0.79**	**0.68**	**0.81**	**0.85**	**0.87**	**0.92**
Context-aware model (SPM)	**0.69**	**0.77**	**0.79**	**0.86**	**0.88**	**0.72**	**0.84**	**0.85**	**0.92**	**0.95**

Random guess is 0.5 (Percentage as observation ratios).
The bold values indicate the best result (i.e., highest accuracy) in each setting.

to be predicted than the cooking activity dataset. This is because the sensor dataset detects actionlets and objects based on sensors, which generate less noise and can mitigate occlusions.

5 Summary

In this chapter, we propose a novel approach to model complex temporal composition of actionlets for activity prediction. To build an effective representation for prediction, human activities can be characterized by a complex temporal composition of constituent simple actions and interacting objects. Different from early detection on short-duration simple actions, we propose a novel framework for *long*-duration complex activity prediction by discovering three key aspects of activity: **Causality**, **Context-cue**, and **Predictability**. The major contributions of our work include: (1) a general framework is proposed to systematically address the problem of complex activity prediction by mining temporal sequence patterns; (2) PST is introduced to model causal relationships between constituent actions, where both large and small order Markov dependencies between action units are captured; (3) the context-cue, especially interactive objects information, is modeled through SPM, where a series of action and object co-occurrence are encoded as a complex symbolic sequence; (4) we also present a PAF to depict the predictability of each kind of activity. We have empirically shown that incorporating causality, context-cue, and predictability is particularly beneficial for predicting various kinds of human activity in diverse environments. Our approach is useful for activities with deep hierarchical structure or repetitive structure.

References

1. Abbeel, P., Ng, A.Y.: Apprenticeship learning via inverse reinforcement learning. In: Proceedings of ACM International Conference on Machine Learning, p. 1 (2004)
2. Agrawal, R., Srikant, R.: Mining sequential patterns. In: Proceedings of IEEE International Conference on Data Engineering, pp. 3–14 (1995)
3. Agrawal, R., Srikant, R., et al.: Fast algorithms for mining association rules. In: Proceedings of International Conference on Very Large Data Bases, vol. 1215, pp. 487–499 (1994)
4. Baker, C.L., Saxe, R., Tenenbaum, J.B.: Action understanding as inverse learning. J. Cogn. **113**(3), 329–349 (2009)
5. Begleiter, R., El-Yaniv, R., Yona, G.: On prediction using variable order Markov models. J. Artif. Intell. Res. **22**, 385–421 (2004)
6. Brendel, W., Todorovic, S.: Learning spatiotemporal graphs of human activities. In: Proceedings of IEEE International Conference on Computer Vision, pp. 778–785 (2011)
7. Brendel, W., Fern, A., Todorovic, S.: Probabilistic event logic for interval-based event recognition. In: Proceedings of IEEE Conference on Computer Vision and Pattern Recognition, pp. 3329–3336 (2011)
8. Brown, P.F., Desouza, P.V., et al.: Class-based n-gram models of natural language. J. Comput. Linguist. **18**(4), 467–479 (1992)
9. Cao, Y., Barrett, D., et al.: Recognizing human activities from partially observed videos. In: Proceedings of IEEE Conference on Computer Vision and Pattern Recognition. IEEE (2013)
10. Choi, W., Shahid, K., Savarese, S.: Learning context for collective activity recognition. In: Proceedings of IEEE Conference on Computer Vision and Pattern Recognition, pp. 3273–3280 (2011)
11. Collins, R., Zhou, X., Teh, S.K.: An open source tracking testbed and evaluation web site. In: IEEE International Workshop Performance Evaluation of Tracking and Surveillance (2005)
12. Davis, J.W., Tyagi, A.: Minimal-latency human action recognition using reliable-inference. J. Image Vision Comput. **24**(5), 455–472 (2006)
13. Desobry, F., Davy, M., et al.: An online kernel change detection algorithm. IEEE Trans. Signal Process. **53**(8), 2961–2974 (2005)
14. Dong, G.: Sequence Data Mining. Springer, New York (2009)
15. Facca, F.M., Lanzi, P.L.: Mining interesting knowledge from weblogs: a survey. J. Data Knowl. Eng. **53**(3), 225–241 (2005)
16. Fan, Q., Bobbitt, R., et al.: Recognition of repetitive sequential human activity. In: Proceedings of IEEE Conference on Computer Vision and Pattern Recognition, pp. 943–950 (2009)
17. Gaidon, A., Harchaoui, Z., Schmid, C.: Actom sequence models for efficient action detection. In: Proceedings of IEEE Conference on Computer Vision and Pattern Recognition, pp. 3201–3208 (2011)
18. Gupta, A., Kembhavi, A., Davis, L.S.: Observing human-object interactions: using spatial and functional compatibility for recognition. IEEE Trans. Pattern Anal. Mach. Intell. **31**(10), 1775–1789 (2009)
19. Haider, P., Brefeld, U., Scheffer, T.: Supervised clustering of streaming data for email batch detection. In: Proceedings of ACM International Conference on Machine learning, pp. 345–352 (2007)
20. Hamid, R., Maddi, S., et al.: A novel sequence representation for unsupervised analysis of human activities. J. Artif. Intell. **173**(14), 1221–1244 (2009)
21. Han, D., Bo, L., et al.: Selection and context for action recognition. In: Proceedings of IEEE International Conference on Computer Vision, pp. 1933–1940 (2009)
22. Hoai, M., De la Torre, F.: Max-margin early event detectors. In: Proceedings of IEEE Conference on Computer Vision and Pattern Recognition, pp. 2863–2870 (2012)
23. Ikizler-Cinbis, N., Sclaroff, S.: Object, scene and actions: combining multiple features for human action recognition. In: Proceedings of European Conference on Computer Vision, pp. 494–507 (2010)

24. Ivanov, Y.A., Bobick, A.F.: Recognition of visual activities and interactions by stochastic parsing. IEEE Trans. Pattern Anal. Mach. Intell. **22**(8), 852–872 (2000)
25. Jiang, Y.G., Li, Z., Chang, S.F.: Modeling scene and object contexts for human action retrieval with few examples. IEEE Trans. Circuits Syst. Video Technol. **21**(5), 674–681 (2011)
26. Kim, K.-J.: Financial time series forecasting using support vector machines. J. Neurocomput. **55**(1), 307–319 (2003)
27. Kitani, K.M., Ziebart, B.D., et al.: Activity forecasting. In: Proceedings of European Conference on Computer Vision, pp. 201–214 (2012)
28. Kollar, T., Roy, N.: Utilizing object-object and object-scene context when planning to find things. In: Proceedings of IEEE International Conference on Robotics and Automation, pp. 2168–2173 (2009)
29. Kwak, S., Han, B., Han, J.H.: Scenario-based video event recognition by constraint flow. In: Proceedings of IEEE Conference on Computer Vision and Pattern Recognition, pp. 3345–3352 (2011)
30. Laptev, I.: On space-time interest points. Int. J. Comput. Vis. **64**(2), 107–123 (2005)
31. Levine, S., Popovic, Z., Koltun, V.: Nonlinear inverse reinforcement learning with gaussian processes. In: Proceedings of Neural Information Processing Systems, vol. 24, pp. 1–9 (2011)
32. Li, K., Fu, Y.: Prediction of human activity by discovering temporal sequence patterns. IEEE Trans. Pattern Anal. Mach. Intell. **36**(8), 1644–1657 (2014)
33. Li, K., Hu, J., Fu, Y.: Modeling complex temporal composition of actionlets for activity prediction. In: Proceedings of European Conference on Computer Vision, pp. 286–299 (2012)
34. Mabroukeh, N.R., Ezeife, C.I.: A taxonomy of sequential pattern mining algorithms. ACM J. Comput. Surv. **43**(1), 3 (2010)
35. Marszalek, M., Laptev, I., Schmid, C.: Actions in context. In: Proceedings of IEEE Conference on Computer Vision and Pattern Recognition, pp. 2929–2936 (2009)
36. Munoz, D., Bagnell, J., Hebert, M.: Stacked hierarchical labeling. In: Proceedings of European Conference on Computer Vision, pp. 57–70 (2010)
37. Munoz, D., Bagnell, J.A., Hebert, M.: Co-inference for multi-modal scene analysis. In: Proceedings of European Conference on Computer Vision, pp. 668–681 (2012)
38. Nasraoui, O., Soliman, M., et al.: A web usage mining framework for mining evolving user profiles in dynamic web sites. IEEE Trans. Knowl. Data Eng. **20**(2), 202–215 (2008)
39. Neill, D., Moore, A., Cooper, G.: A Bayesian spatial scan statistic. In: Proceedings of Neural Information Processing Systems, vol. 18, pp. 1003–1010 (2006)
40. Niebles, J.C., Wang, H., Fei-Fei, L.: Unsupervised learning of human action categories using spatial-temporal words. Int. J. Comput. Vis. **79**(3), 299–318 (2008)
41. Niebles, J., Chen, C.-W., Fei-Fei, L.: Modeling temporal structure of decomposable motion segments for activity classification. In: Proceedings of European Conference on Computer Vision, pp. 392–405 (2010)
42. Pei, M., Jia, Y., Zhu, S.-C.: Parsing video events with goal inference and intent prediction. In: Proceedings of IEEE International Conference on Computer Vision, pp. 487–494 (2011)
43. Roggen, D., Calatroni, A., et al.: Collecting complex activity data sets in highly rich networked sensor environments. In: Proceedings of International Conference on Networked Sensing Systems, pp. 233–240 (2010)
44. Rohrbach, M., Amin, S., et al.: A database for fine grained activity detection of cooking activities. In: Proceedings of IEEE Conference on Computer Vision and Pattern Recognition, pp. 1194–1201 (2012)
45. Ron, D., Singer, Y., Tishby, N.: The power of amnesia: learning probabilistic automata with variable memory length. J. Mach. Learn. **25**(2), 117–149 (1996)
46. Ryoo, M.S.: Human activity prediction: early recognition of ongoing activities from streaming videos. In: Proceedings of IEEE International Conference on Computer Vision, pp. 1036–1043 (2011)
47. Ryoo, M.S., Aggarwal, J.K.: Recognition of composite human activities through context-free grammar based representation. In: Proceedings of IEEE Conference on Computer Vision and Pattern Recognition, pp. 1709–1718 (2006)

48. Si, Z., Pei, M., et al.: Unsupervised learning of event and-or grammar and semantics from video. In: Proceedings of IEEE International Conference on Computer Vision, pp. 41–48 (2011)
49. Smith, T.F., Waterman, M.S.: Identification of common molecular subsequences. J. Mol. Biol. **147**, 195–197 (1981)
50. Srivastava, J., Cooley, R., et al.: Web usage mining: discovery and applications of usage patterns from web data. ACM SIGKDD Explorations Newsl. **1**(2), 12–23 (2000)
51. Turaga, P.K., Veeraraghavan, A., Chellappa, R.: From videos to verbs: mining videos for activities using a cascade of dynamical systems. In: Proceedings of IEEE Conference on Computer Vision and Pattern Recognition, pp. 1–8 (2007)
52. Wang, K., Xu, Y., Yu, J.X.: Scalable sequential pattern mining for biological sequences. In: Proceedings of ACM International Conference on Information and Knowledge Management, pp. 178–187 (2004)
53. Wang, H., Klaser, A., et al.: Action recognition by dense trajectories. In: Proceedings of IEEE Conference on Computer Vision and Pattern Recognition, pp. 3169–3176 (2011)
54. Wang, J., Liu, Z., Wu, Y., Yuan, J.: Mining actionlet ensemble for action recognition with depth cameras. In: Proceedings of IEEE Conference on Computer Vision and Pattern Recognition, pp. 1290–1297. IEEE (2012)
55. Yao, B., Fei-Fei, L.: Modeling mutual context of object and human pose in human-object interaction activities. In: Proceedings of IEEE Conference on Computer Vision and Pattern Recognition, pp. 17–24 (2010)
56. Zhao, Q., Bhowmick, S.S.: Sequential pattern mining: a survey. Technical Report CAIS Nanyang Technological University, Singapore, pp. 1–26 (2003)
57. Ziebart, B.D., Maas, A., et al.: Maximum entropy inverse reinforcement learning. In: Proceedings of AAAI, pp. 1433–1438 (2008)
58. Ziebart, B.D., Ratliff, N., et al.: Planning-based prediction for pedestrians. In: Proceedings of IEEE International Conference on Intelligent Robots and Systems, pp. 3931–3936 (2009)

Chapter 8
Time Series Modeling for Activity Prediction

Kang Li, Sheng Li, and Yun Fu

1 Introduction

Human Motion Analysis (HMA) is a highly interdisciplinary research area which attracts great interests from computer vision, machine learning, multimedia, and medical research communities, due to the potential applications ranging from, human–computer interaction, security (intelligent surveillance), health (assistive clinical studies), information technology (content-based video retrieval), entertainments (special effects in film production and somatosensory game) to all aspects of our daily life. As an important research thread of HMA, action recognition builds the basis for all of the above-mentioned applications.

Though action recognition has been well studied in the literature, activity prediction is quite a new topic for the fields of multimedia and computer vision. Spatial and temporal structures of activities bring new challenges as well as new opportunities

©{Kang Li, Sheng Li, and Yun Fu | IEEE} {2014}. This is a minor revision of the work published in {Proceedings of the Data Mining (ICDM), 2014 IEEE International Conference on, pp. 310–319.}, http://dx.doi.org/10.1109/ICDM.2014.100.

K. Li (✉) • S. Li
Department of Electrical and Computer Engineering, Northeastern University,
360 Huntington Ave. Boston, MA 02115, USA
e-mail: li.ka@husky.neu.edu; kongkong115@gmail.com; shengli@ece.neu.edu

Y. Fu
Department of Electrical and Computer Engineering and College of Computer and Information Science (Affiliated), Northeastern University, 360 Huntington Ave. Boston, MA 02115, USA
e-mail: yunfu@ece.neu.edu

© Springer International Publishing Switzerland 2016
Y. Fu (ed.), *Human Activity Recognition and Prediction*,
DOI 10.1007/978-3-319-27004-3_8

for the research community. This chapter focuses on early recognition[1] of ongoing activities, which is beneficial for a large variety of time-critical scenarios. For example, in human–computer interaction, people's intension can be predicted by early recognizing human actions captured by sensors or depth cameras, which may greatly reduce the system response time and provide a more natural experience of communication. In many real-time somatosensory games, early recognition of human actions can reduce the sense of delay and create richer, more enjoyable gaming experiences.

In this chapter, we introduce a novel approach to early classify human activities represented by multivariate time series (m.t.s.) data [23], where the spatial structure of activities is encoded by the dimensions of predefined human body model, and the temporal structure of activities is modeled by two types of time pattern: (1) **Temporal Dynamics** and (2) **Sequential Cue**, shown in Fig. 8.1.

Our key idea is that m.t.s. activity observation can be modeled as an instantiation of a Multivariate Marked Point-Process (Multi-MPP). Each dimension of Multi-MPP characterizes the *temporal dynamics* of a particular body part of the activity, where both timing and strength information are kept. Since a full parameter estimation of Multi-MPP can easily become impractical with the increase of the number of time instants, we introduce Multilevel-Discretized Marked Point-Process Model (MD-MPP), which is a class of Multi-MPP that can ensure a good piece-wise stationary property both in time-domain and mark-space while keeping

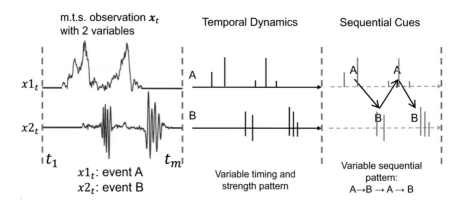

Fig. 8.1 Temporal Dynamics and Sequential Cues. Each dimension of x_t is considered as an event, whose timing and strength information are characterized by MPP. The temporal order of firing patterns among events contains important sequential cue to early recognize the class of ongoing observation. For example, assuming $A \rightarrow B$ pattern is discriminative for this class, then we can make classification decision when we observed only half of the m.t.s. data

---

[1]In our discussion, we use "early recognition," "early classification," or "prediction" interchangeably to refer to the same learning task: "identifying the class label of human activities with partial observation of temporally incomplete executions."

dynamics as much as possible. Based on this representation, another important temporal pattern of early classification, *sequential cue*, becomes formalizable. We construct a probabilistic suffix tree (PST) to represent sequential patterns among variables (feature dimensions) in terms of variable order Markov dependencies. We use MD-MPP+TD to denote this extended version of our approach, in which *temporal dynamics* and *sequential cue* are integrated. In order to test the efficacy of our method, comprehensive evaluations are performed on two real-world human activity datasets. The proposed algorithms achieve superior performance for activity prediction with m.t.s. observation.

2   Related Work

In general, there are four categories of works that are related to ours.

Action/Activity Recognition   A large number of methods have been proposed for recognizing human actions. Here we focus on methods most related to 3D actions, where observations captured both spatial and temporal information of activity. Readers interested in 2D action can refer to some recent survey [1] on this topic. Most of the existing work [20, 26, 27, 35] on 3D action recognition are extensions from 2D case by either customizing the features to depth camera or adjusting 2D action recognition algorithms so that it can handle new challenges introduced by the depth sensor. Xia, Aggarwal [35] extends the classic 2D action feature to 3D counterpart, Depth STIP, which is basically a filtering method to detect interest points from RGB-D videos with noise reduction. Wang et al. [32] proposed to represent 3D actions as a set of selected joints which are considered more relevant and informative to the task. And they use a framework of multiple-kernel SVM, where each kernel corresponds to an informative joint. Also, for a more standard and comprehensive evaluation of this particular task, new dataset is also provided recently [11].

Action/Activity Prediction   Action/Activity Prediction is quite a new topic itself. Only a few existing works specifically focus on this task, but limited to 2D. The work of [31] first argued that the goal of activity prediction is to recognize unfinished single actions from observation of its early stage. Two extensions of bag-of-words paradigm, dynamic BoW and integral BoW are proposed to handle the sequential nature of human activities. The work of [4] extended [31] to recognize human actions from partially observed videos, where an unobserved subsequence may occur at any time by yielding a temporal gap in the video. The work of [19] proposed a discriminative model to enforce the label consistency between segments. The work of [12] proposed a max-margin framework for early event detection, in which video frames are simulated as sequential event streams. To implement the idea of action prediction for long duration, more complex human activities, [21, 22] introduce the concept of actionlets, where sequential nature of action units is explored for the purpose of recognizing the activity class as early as possible.

Time Series Early Classification While there is a vast amount of literature on classification of time series (see reviews [8, 17], and recent work [7, 15, 33, 34, 38, 39]), early classification of ongoing time series is ignored until quite recently [9, 36, 37]. The unique and non-trivial challenge here is that either features or distance metrics formulated in previous work for classification of time series might not be robust, when whole time series is not available. Additionally, early classification always makes stricter demands on time efficiency, because the algorithm will lose its merit, if it unintentionally forces us to wait till the end of time series. To the best of our knowledge, the work of [37] first explicitly proposed a solution of early classification of time series to the community, though similar concepts have been raised in other two works [3, 29]. They developed ECTS (Early Classification on Time Series) algorithm, which is an extension of 1NN classification method. ECTS evaluates neighbors both in full observation and prefixes of time series. But their algorithm is only limited to u.t.s. data and assuming that all time series samples have the same length. Following the spirit of the classic work in [38] on discovering interpretable time series shapelets, [9] and [36] extend it to the early classification scenarios. However all three methods are distance based approaches, the inherent efficiency problem is not considered for earliness.

Though the problem of early classification arises in a wide variety of applications, it is quite a new topic for the domain of statistical learning. Existing works are either focusing on Univariate Time Series (u.t.s.) [36, 37] or from application perspectives by tuning on traditional time series classification models [9]. The disadvantages of previous work are three folds. First, many approaches assume that the time series observations from the same class will always have equal durations, which reduced the problem into a significantly simplified one (simple distance measuring between samples). In terms of early classification task, the equal length assumption also implicitly means that we can exactly tell how much an ongoing time series has progressed and when it will be finished. But in most of the real-world applications, this assumption cannot hold. Second, an important factor, temporal correlations among variables of m.t.s. are not fully considered, which can be quite informative for identifying the object class at early stage of observation. For instance, in human action recognition, a particular action is a combined motion of multiple joints with temporal order. Third, all previous work [9, 36, 37] are extensions of traditional distance based approach, which are computationally too demanding. However, in many cases, the practical merit of early classification lies in a quick and accurate recognition.

Point Process in Vision As a special type of stochastic process, point process has gained a lot of attention recently in the statistical learning community because of its powerful capability on modeling and analyzing rich dynamical phenomena [10, 14, 18, 28]. Adopting a point process representation of random events in time opens up pattern recognition to a large class of statistical models that have seen wide applications in many fields.

3   Preliminaries

3.1   Notation and Problem Definition

For better illustration, Table 8.1 summarizes the abbreviations and notations used throughout the chapter. We begin by defining the key terms in the chapter. We use lowercase letters to represent scalar values, lowercase bold letters to represent vectors. We use uppercase letters to represent time series, uppercase bold letters to represent sets.

Definition 1. *Multivariate Time Series*: A multivariate time series $X = \{\mathbf{x}_t : t \in \mathbf{T}\}$ is an ordered set of real-valued observations, where $\mathbf{T}$ is the index set consisting of all possible time stamps. If $\mathbf{x}_t \in \mathbb{R}^d$, where $d > 1$, for instance $\mathbf{x}_t = \langle x_t^1, x_t^2, \ldots, x_t^d \rangle$, X is called a d-dimensional m.t.s..

Table 8.1 Symbol and abbreviation

Abbreviation	Description
u.t.s	Univariate time series
m.t.s	Multivariate time series
MPP	Marked point process
MD-MPP	Multilevel-discretized marked point-process
1NN	1 Nearest neighbor
DTW	Dynamic time warping
PST	Probabilistic suffix tree
VMM	Variable order Markov model
Symbol	*Description*
X	Observation of time series with full length
X', Y'	Ongoing time series
$\mathbf{X}^d$	Set of d-dimensional m.t.s.
$\mathbf{D}$	Time series training dataset
$\mathbf{T}$	Time (index set)
$\mathbf{C}$	Set of class labels
$|X|$	Length of time series
$\mathcal{F}$	Classifier
$\tilde{\mathbf{N}}$	Multivariate point-process
$\ddot{\mathbf{N}}$	Multivariate marked point-process
S	Number of segments by factoring time line
Λ	Trained MD-MPP model
$\mathbf{E}$	Set of events
$\overline{\mathbf{D}}_\Lambda$	Set of sampled discrete event streams from model Λ
$\overline{\mathbf{D}}_{Y'}$	Set of sampled discrete event streams from testing Y'
$\bar{a}_i$	Discrete event stream

In this chapter, observations $\mathbf{x}_t$ are always arranged by temporal order with equal time intervals.

Definition 2. *Classification of Multivariate Time Series*: An m.t.s. $X = \{\mathbf{x}_t : t \in \mathbf{T}\}$ may globally carry a class label. Given $\mathbf{C}$ as a set of class labels, and a training set $\mathbf{D} = \{\langle X_i, C_i \rangle : C_i \in \mathbf{C}, i = 1, \ldots, n\}$, the task of classification of m.t.s. is to learn a classifier, which is a function $\mathcal{F} : \mathbf{X}^d \rightarrow \mathbf{C}$, where $\mathbf{X}^d$ is the set of d-dimensional m.t.s..

We use $|X|$ to represent the *length* of time series, namely $X = \{\mathbf{x}_{t_1}, \mathbf{x}_{t_2}, \ldots, \mathbf{x}_{t_{|X|}}\}$. By default, X is considered as the full-length observed time series, while a corresponding *ongoing time series* of X is denoted as $X' = \{\mathbf{x}'_{t_1}, \mathbf{x}'_{t_2}, \ldots, \mathbf{x}'_{t_{|X'|}}\}$, where $\mathbf{x}'_{t_i} = \mathbf{x}_{t_i}$ for $i = 1, \ldots, |X'|$, and $t_{|X'|} < t_{|X|}$. The ratio $p = |X'|/|X|$ is called the *progress level* of X'. It's obvious that the progress level of full-length observed time series is always 1. We use X'_p to indicate an ongoing time series with progress level p.

Definition 3. *Early Classification of Multivariate Time Series*: Given training set $\mathbf{D} = \{\langle X_j, C_j \rangle : C_j \in \mathbf{C}, j = 1, \ldots, n\}$ with n m.t.s. samples, the task of early classification of m.t.s. is to learn a classifier, which is a function $\mathcal{F} : \mathbf{X}' \rightarrow \mathbf{C}$, where $\mathbf{X}'$ is the set of ongoing m.t.s..

Specifically, we can do classification along the progress of time series, and predict the class label at different progress levels of X, generating a bunch of decisions, $\{\mathcal{F}(X'_{p_1}), \mathcal{F}(X'_{p_2}), \ldots, \mathcal{F}(X'_1)\}$. In this chapter we use 5 % of full duration as an interval of generating a new prediction result, which results in 20 rounds of classification for different progress levels. Intuitively, the prediction accuracy should go up with the increasing progress level, since we observed more information. But, interestingly, through our evaluations at later sections, we found that, sometimes, it is quite contradictory to our common sense. The reason is that observations at different segments of time series may have different discriminativeness for classification task, and how the discriminative segments distribute along the timeline really depends on the data.

3.2 Multivariate Marked Point-Process

In probability theory, *stochastic process* is sequence of random variables indexed by a totally ordered set $\mathbf{T}$ ("time"). *Point process* is a special type of stochastic process which is frequently used as models for firing pattern of random events in time. Specifically, the process counts the number of events and records the time that these events occur in a given observation time interval.

Definition 4. A d-dimensional *multivariate point-process* is described by $\tilde{\mathbf{N}} = \langle N^1, N^2, \ldots, N^d \rangle$, where $N^i = \{t^i_1, t^i_2, \ldots, t^i_m\}$ is a univariate point-process, and t^i_k

indicates the time stamps on which a particular "event" or "property"[2] x_i has been detected. $N^i(t)$ is the total number of observed event x_i in the interval $(0, t]$, for instance, $N^i(t^i_k) = k$. Then, $N^i(t + \Delta t) - N^i(t)$ represents the number of detections in the small region Δt. Similarly, $\tilde{\mathbf{N}}(t) = \langle N^1(t), N^2(t), \dots, N^d(t) \rangle$,

By letting $\Delta t \to 0$, we can have the *intensity function* $\mathbf{\Lambda}(t) = \{\lambda^i(t)\}$, which indicates the expected occurrence rate of the event x^i at time t: $\lambda^i(t) = \lim_{\Delta t \to 0} N^i(t + \Delta t) - N^i(t)$ [6]. This is the key to identify a point process.

In many real world applications, the time landmarks of events arise not as the only object of study but as a component of a more complex model, where each landmark is associated with other random elements $M^i = \{x^i_1, x^i_2, \dots, x^i_m\}$, called marks, containing further information about the events. Each (t^i_k, x^i_k) is a marked point, and the sequence $\{(t^i_k, x^i_k)\}$ of marked points is referred to as a *marked point processes*.

Definition 5. A d-dimensional *multivariate marked point process* is described as follows:

$$\ddot{\mathbf{N}} = \langle \{N^1, M^1\}, \{N^2, M^2\}, \dots, \{N^d, M^d\} \rangle \tag{8.1}$$

where $\{N^i, M^i\} = \{(t^i_k, x^i_k)\}$ on $\mathbb{R}^+ \times \mathbb{R}$ is a univariate marked point process.

4   Methodology

In this section, we describe the proposed early classification approach for m.t.s. data. Our basic idea is to construct early classification function $\mathcal{F}(Y')$ by using the knowledge learned from a *temporal dynamics* model $\Pr(Y'|\mathbf{\Lambda})$ (Sect. 4.1) and a *sequential cue* model $\Pr(Y'|\mathbf{\Phi})$ (4.2). We use MD-MPP to denote the first model, and MD-MPP+TD to denote the second model. Given an ongoing m.t.s. Y' in a domain application with $|\mathbf{C}|$ classes, the final prediction functions can be written as:

$$\text{MD-MPP:} \qquad \mathcal{F}(Y') = \arg\max_{c \in \mathbf{C}} \{\Pr^c(Y'|\mathbf{\Lambda})\};$$

$$\text{MD-MPP+TD:} \qquad \mathcal{F}(Y') = \arg\max_{c \in \mathbf{C}} \{\Pr^c(Y'|\mathbf{\Phi})\}.$$

The bases of our method are the following two insights: (1) m.t.s. can be interpreted as an instantiation of a Multi-MPP. Each dimension of Multi-MPP characterizes the temporal dynamics of a particular property of the object, where both timing and strength information are kept; (2) The identification of sequential patterns among multiple variables of m.t.s. allows us to utilize these sequential cues for early classification.

---

[2]In this chapter, the concepts "variable," "property," or "event" are interchangeably used to refer to a certain dimension of m.t.s.

Specifically, our approach consists of two stages. The first stage encodes the m.t.s. as a multi-level discretized marked point process, which not only characterizes the temporal dynamics of m.t.s., but also provides discretized intensity map that governs the generation of discrete events streams. The second stage analyzes these discrete events streams to discover the temporal correlations. We will go to the details of each component of our approach in the following subsections.

4.1 Temporal Dynamics

In this chapter, we focus on multivariate time series data. In our opinion, these observations can be (1) adequately represented as a collection of perceptual events that are tied to a common clock or constant frame rate, and (2) decoded according to the temporal statistics of such events. The need therefore arises to formulate and evaluate recognition strategies that can operate on representations based on the firing patterns of various events. In the following, we will introduce our multilevel-discretized marked point-process (MD-MPP) model $\ddot{\mathbf{N}}_X$ to achieve this.

Given a d-dimensional m.t.s. $X = \{\mathbf{x}_{t_1}, \mathbf{x}_{t_2}, \dots, \mathbf{x}_{t_{|X|}}\}$, where $\mathbf{x}_t = \langle x_t^1, x_t^2, \dots, x_t^d \rangle$, and $t = t_1, t_2, \dots, t_{|X|}$. We consider each dimension x^i as a noisy detector of certain perceptual event. Those event detectors generate continuous values which indicate the strength or the confidence about the detection. We call these continuous value based observations as *marks*. Then the corresponding marked point process representation of X is:

$$\ddot{\mathbf{N}}_X = \left\langle \{N_X, M_X^1\}, \{N_X, M_X^2\}, \dots, \{N_X, M_X^d\} \right\rangle, \tag{8.2}$$

where $\{N_X, M_X^i\} = \{(t_k, x_{t_k}^i)\}$, and $k = 1, \dots, |X|$. We can see that different variables shared a common clock N_X.

To allow for more clarity in understanding the approach, we will develop the model step by step by relaxing assumptions from ideal case to the real case. In computation dealing with time series, the number of time instants is often in the hundreds or thousands. Dealing with so many variable density function is cumbersome at best and often impractical. We need to think about special cases which may simplify things. The first drastic simplification is to have an ideal case with the following assumptions:

Assumption 1: For each event x^i, the corresponding point process is an independent stochastic sequence (to be relaxed in Sect. 4.1);

Assumption 2: We have a perfect detector for each event x^i, namely, $x_t^i \in \{0, 1\}$, where a spiking ($x_t^i = 1$) indicates occurrence of x^i, or there is no detection of x^i ($x_t^i = 0$) (to be relaxed in Sect. 4.1);

Assumption 3: For m.t.s. X, events are independent from each other (to be relaxed in Sect. 4.2).

Based on above assumptions, we have first representation model for m.t.s.: *Stationary point process*:

$$\Pr(N^i) = \prod_{k=1}^{|X|} \frac{(\lambda^i \Delta t)^{\mathbf{1}_{N^i}(t_k)}}{\mathbf{1}_{N^i}(t_k)!} e^{-\lambda^i \Delta t} \tag{8.3}$$

$$= (\lambda^i \Delta t)^{m^i} e^{-\lambda^i T}$$

where X is an m.t.s., $N^i = \{t_k | x_{t_k}^i = 1, t_k \le t_{|X|}\}$ is the point process for event x^i of X. $N^i(t_{|X|}) = |\{t_k | x_{t_k}^i = 1, t_k \le t_{|X|}\}| = m^i$ is the numbers of detection of event x^i in X, $\Delta t = t_{k+1} - t_k$ is the time interval between two consecutive observations. Assuming the whole process is contained in interval $(0, T]$, then $T = (|X| - 1)\Delta t$. The indicator function $\mathbf{1}_{N^i}(t_k)$ is 1 if $t_k \in N^i$ and 0 otherwise.

Given training dataset $\mathbf{D} = \{\langle X_j, C_j \rangle : C_j \in \mathbf{C}, j = 1, \ldots, n\}$, and point process representation $\tilde{\mathbf{N}} = \langle N^1, N^2, \ldots, N^d \rangle$ and duration time T, the data likelihood can be computed as

$$\Pr(\tilde{\mathbf{N}}|\mathbf{D}) = \prod_{i=1}^{d} \Pr(N^i|\mathbf{D}) = \prod_{i=1}^{d} (\lambda^i(\mathbf{D})\Delta t)^{m^i} e^{-\lambda^i(\mathbf{D})T} \tag{8.4}$$

where $\lambda^i(\mathbf{D})$ depends both on the event and the training data.

Then training this model amounts to estimating $\lambda^i(\mathbf{D})$ for each $\langle i, \mathbf{D} \rangle$ pair. If we are given n training sequences containing in $\mathbf{D}$, and there are m_j^i of landmarks (spiking) of event x^i in j-th training sample, then, we can estimate $\lambda^i(\mathbf{D})$ by using the maximum log-likelihood estimation, which is:

$$\lambda^{i*}(\mathbf{D}) = \arg\max_{\lambda^i} \log(\lambda^i(\mathbf{D})\Delta t)^{m^i} e^{-\lambda^i(\mathbf{D})T}) \tag{8.5}$$

$$= \frac{\Sigma_{j=1}^n m_j^i}{\Sigma_{j=1}^n \Delta t |X_j|}.$$

Relax Assumption 1 Next, we will relax assumption 1 by adding *dynamic* property in the point process representation. This follows the *piece-wise stationary global-wise dynamic point process*:

$$\Pr(N^i) = \prod_{s=1}^{S} \frac{(\lambda^i(s)\Delta t \Delta \tau)^{m_s^i}}{m_s^i!} e^{-\lambda^i(s)\Delta t \Delta \tau} \tag{8.6}$$

where assuming we evenly divide the time line into S pieces of equal length segments. Inside each segment, the point process is assumed stationary. $\Delta \tau = \lfloor |X|/S \rfloor$ is the division length in terms of number of observations, so the progress level at the end of s-th time division is $p = (s\Delta\tau\Delta t)/(|X|\Delta t) = s\Delta\tau/|X|$. m_s^i is the number of detection of event x^i in time division s.

Given training m.t.s. dataset $\mathbf{D}$, and point process representation $\tilde{\mathbf{N}}$, the data likelihood can be computed as

$$\Pr(\tilde{\mathbf{N}}|\mathbf{D}) = \prod_{i=1}^{d} \Pr(N^i|\mathbf{D})$$

$$= \prod_{i=1}^{d}\prod_{s=1}^{S} \frac{(\lambda^i(s,\mathbf{D})\Delta t\Delta\tau)^{m_s^i}}{m_s^i!} e^{-\lambda^i(s,\mathbf{D})\Delta t\Delta\tau} \tag{8.7}$$

where the intensity function $\lambda^i(s,\mathbf{D})$ depends on the event, the time division (progress level) and the training data.

Then training this model amounts to estimating $\lambda^i(\mathbf{D})$ for each tuple $\langle i, s, \mathbf{D}\rangle$. If we are given n training sequences containing in $\mathbf{D}$, and there are $m_{j,s}^i$ of landmarks (spiking) of event x^i in j-th training sample's s-th division, then, we can estimate $\lambda^i(s,\mathbf{D})$ by using the maximum log-likelihood estimation, which is:

$$\lambda^{i*}(s,\mathbf{D}) = \frac{\sum_{j=1}^{n} m_{j,s}^i}{\sum_{j=1}^{n} \Delta t\Delta\tau_j}. \tag{8.8}$$

Relax Assumption 2 In practice, we always get a noisy detector for each event, such as m.t.s. data. In the following, we will keep piece-wise stationary property and relax assumption 2 by allowing event detectors generating continuous values which may indicate the strength or the confidence about the detection. We call these continuous value based observations as *marks*, then the whole m.t.s. can be extended to a marked point process representation. To deal with this complexity, we introduce the *multilevel-discretized marked point-process* (MD-MPP).

In this case, intensity parameter λ will depend on both time and mark. In this chapter, we assume all feature dimensions have been normalized to $[0, 1]$ respectively, which results in the mark space within $[0, 1]$. We build a multi-level discretization of mark space by splitting it into L levels. Then the point process factors into L levels of independent processes operating in each level of the mark space for a particular event.

$$\Pr(\{N_X, M_X^i\}) \tag{8.9}$$

$$= \prod_{l=1}^{L}\prod_{s=1}^{S} \frac{(\lambda^i(s,l)\Delta t\Delta\tau)^{m_{s,l}^i}}{m_{s,l}^i!} e^{-\lambda^i(s,l)\Delta t\Delta\tau}$$

Given training m.t.s. dataset $\mathbf{D}$, and multivariate marked point process representation $\tilde{\mathbf{N}}$, the data likelihood can be computed as

$$\text{Pr}(\ddot{\mathbf{N}}|\mathbf{D}) = \prod_{i=1}^{d} \text{Pr}(\{N^i, M^i\}|\mathbf{D})$$

$$= \prod_{i=1}^{d} \prod_{l=1}^{L} \prod_{s=1}^{S} \frac{(\lambda^i(s,l,\mathbf{D})\Delta t \Delta \tau)^{m^i_{s,l}}}{m^i_{s,l}!} e^{-\lambda^i(s,l,\mathbf{D})\Delta t \Delta \tau} \quad (8.10)$$

where the intensity function $\lambda^i(s,l,\mathbf{D})$ depends on the event, the time division (progress level), the mark space level, and the training data. Now, we can formalize our two key steps in early classification.

1. Learning MD-MPP Given n training samples, the maximum log-likelihood estimation of model parameters is:

$$\lambda^{i^*}(s,l,\mathbf{D}) = \frac{\sum_{j=1}^{n} m^i_{j,s,l}}{\sum_{j=1}^{n} \Delta t \Delta \tau_j}. \quad (8.11)$$

where $m^i_{j,s,l}$ is the number of landmarks of event x^i in j-th training sample's s-th time division, l-th level of mark space.

2. Early Classification Given an ongoing testing m.t.s Y', and a trained model $\Lambda = \{\lambda_{i,s,l}|L, S, \mathbf{D}\}$ (for simplicity, we use $\lambda_{i,s,l}$ to represent $\lambda^{i^*}(s,l,\mathbf{D})$). First, we construct a structure of Y' by factoring it over time line and mark space in the same way as trained model, so that dynamics can be matched. Then, the likelihood of Y' is:

$$\text{Pr}(Y'|\Lambda) \propto \prod_{i=1}^{d} \prod_{l=1}^{L} \prod_{s=1}^{\lceil p^*S \rceil} (\lambda_{i,s,l}\Delta \tau^*)^{m^i_{s,l}} e^{-\lambda_{i,s,l}\Delta \tau^*} \quad (8.12)$$

where $p^* = |Y'|/(\Delta \tau^* S)$ is our best guessed progress level of Y'. Since the length of m.t.s. can be different, given an ongoing testing m.t.s., we may not know when it will be finished. Therefore, we need to 'guess' the 'right' progress level of it first. Then we can apply our model appropriately. This is an important merit of our approach. Algorithm 1 shows the detail of how we compute p^*.

4.2 Sequential Cue (Relax Assumption 3)

Although MD-MPP provides a good modeling of temporal dynamics for m.t.s., the unrealistic independency assumption between events (Assumption 3) is not relaxed yet. In real applications, m.t.s. always has strong correlations among variables. For instance, in the execution of a particular human action, a few joints will change their angles immediately after other few joints rotated to some degree according to the underlying cognitive "recipe" of that action. The identification of temporal

---

Algorithm 1 Guess the progress level p^*

---

1. **Find the possible range from training set**: Let $\Delta\tau_{min} = \min\{|X_j|/S : j \in \{1,\dots,n\}\}$, $\Delta\tau_{max} = \max\{|X_j|/S : j \in \{1,\dots,n\}\}$, Then, $\tau_D = [\Delta\tau_{min}, \Delta\tau_{max}]$.
2. **Determine the minimum number of segments**: $S' = \min\{\lceil|Y'|/\Delta\tau\rceil : \Delta\tau \in \tau_D\}$, which ensures different guesses of $\Delta\tau$ will be evaluated with the same number of segments, so that the likelihoods computed in step 3 will be comparable.
3. **Evaluate the likelihood**:

$$\Delta\tau^* = \arg\max_{\Delta\tau \in \tau_D} \prod_{i=1}^{d} \prod_{l=1}^{L} \prod_{s=1}^{S'} \frac{(\lambda_{i,s,l}\Delta\tau)^{m_{s,l}^i(\Delta\tau)}}{e^{\lambda_{i,s,l}\Delta\tau}}$$

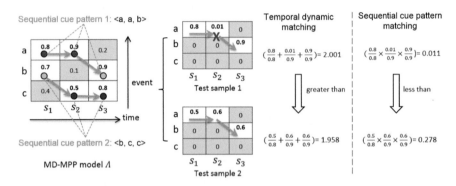

Fig. 8.2 Illustrations of Temporal Dynamics and Sequential Cue, based on our MD-MPP representation. Numbers in the table constitute the model parameter, where each value indicates the firing rate of a certain event within a particular time division. Based on a trained model Λ, we can sample event streams according to these firing rates. For example, event stream $\langle a, a, b\rangle$ and $\langle b, c, c\rangle$ will have more chance to be sampled. With sufficient sampling, these type of Sequential Cue patterns can be mined by using Algorithm 2

correlations among events allows us to utilize these sequential patterns for early classification, which improves the reasoning capability of our model. As illustrated in Fig. 8.2, if we only consider temporal dynamics, test sample 1 will have a better match with the point process model. But in terms of sequential cue patterns, test sample 2 results in a significantly better match. As a complement of MD-MPP model, we introduce the *sequential cue* component of our approach.

Our basic idea is to formalize the notion of *sequential cue* by discovering sequential structure that comes from the order in which events (m.t.s. variables) occur over time. So we need to generate representative discrete event streams by quantizing continuous m.t.s. observations. Since MD-MPP characterizes the rate of occurrence (intensity function) of events at different time division (segments), then we can easily sample event from each segment according to this rate, which results in a discrete event stream. If we generate sufficient number of sample sequences, then the sequential cue patterns will be preserved in the sampling set. Figure 8.2 gives an example showing how we sampled these representative sequences of events

---

Algorithm 2 Construction of Sequential Cue model Φ (*O*-order bounded PST)

1. **Sampling event streams set**: Let $\overline{\mathbf{D}}_\Lambda = \{\bar{a}_1, \ldots, \bar{a}_v\}$ be the training set for learning model Φ.
2. **Creating candidate set U**: Assume h is a subsequence of $\bar{a}_r (r = 1, \ldots, v)$. If $|h| < O$ and $\Pr(h) > \eta$, then put h in **U**. η is a user specified minimal probability requirement for an eligible candidate. $\Pr(h)$ is computed from frequency count.
3. **Testing every candidate h** $\in$ **U**: For any $h \in$ **U**, test following two conditions:

   - (1) $\Pr(e|h) \geq \alpha$, which means the context subsequence h is meaningful for some event e. Here, α is defined by user to threshold a conditional appearance.
   - (2) $\frac{\Pr(e|h)}{\Pr(e|\text{suf}(h))} \geq \beta$, or $\leq 1/\beta$, which means the context h provides extra information in predicting e relative to its longest suffix $\text{suf}(h)$. β is a user specified threshold to measure the difference.
   - **Then**, if h passes above two tests, add h and its suffixes into **U**.

4. **Smoothing the probability to obtain** $\phi(e|h)$:
   For each h in **U**, if $\Pr(e|h) = 0$, we assign a minimum probability γ. In general, the *next event probability function* can be written as:
   $\phi(e|h) = (1 - |\mathbf{E}|\gamma)\Pr(e|h) + \gamma$. Here, γ is the smoothing factor defined empirically.

---

from our trained MD-MPP model. Then the task of finding temporal correlations among features (events) becomes a problem of mining sequential patterns.

Specifically, let $\mathbf{E} = \{e^i : i = 1, \ldots, d \times L\}$ be the set of events.[3] And $\overline{\mathbf{D}}_\Lambda = \{\bar{a}_1, \ldots, \bar{a}_v\}$ consists of v times sampling according to Λ. For instance, $\bar{a}_r = \{e^r_s\}^S_{s=1}, r \in \{1, \ldots, v\}$ is a sampled *event stream*, which means at the j-th segment, we sampled one event $e^r_s \in \mathbf{E}$. We can notice that $\bar{a}_i \in \mathbf{E}^*, |\bar{a}_i| = S$. Specific sampling probability of each event at a particular time (segment) can be computed according to:

$$\Pr_{\text{sample}}(\text{event} = e|\text{segment} = s) = \frac{\lambda_{e,s}}{\sum_{e' \in \mathbf{E}} \lambda_{e',s}} \qquad (8.13)$$

Now the goal is to learn a model $\Phi = \{\phi(e|h) : h \in \mathbf{E}^*, e \in \mathbf{E}\}$, which associates a history h with next possible event e. We call function $\phi(e|h)$ the *next event probability function*. If we define the *history* at j-th time segment of event stream $\bar{a}^i$ as the subsequence $h_j(\bar{a}^i) = \{e^i_j | j \leq S\}$, then the log-likelihood of event stream $\bar{a}^i$, given a Sequential Cue model Φ, can be written as:

$$\Pr(\bar{a}^i|\Phi) = \sum_{j=1}^{S} \log \phi(e^i_j|h_{j-1}(\bar{a}^i)) \qquad (8.14)$$

---

[3]With multilevel-discretized representation, the total number of events becomes $d \times L$. The MD-MPP model can be rewritten as $\Lambda = \{\lambda_{e,s}|e \in \mathbf{E}, s \in \{1, \ldots, S\}\}$ for convenience.

Given an ongoing testing m.t.s. Y', and a trained model $\Phi = \{\phi(e|h)|\overline{D}_A\}$. First, we construct a structure of Y' by factoring it over time line and mark space. We use $\Delta\tau^*$ as the segment length of Y', then we have $S^* = \lceil|Y'|/\Delta\tau\rceil$ segments. Similar to the training process of Λ, we can build an MD-MPP representation for Y' by itself, $\Lambda_{Y'} = \{\lambda'_{i,s,l}|L, S^*, Y'\}$, from which a set of w representative event streams of Y', $\overline{D}_{\Lambda_{Y'}} = \{\bar{b}_1, \ldots, \bar{b}_w\}$, can be sampled in the same way. Then, the likelihood of Y' is:

$$\Pr(Y'|\Phi) \propto \sum_{i=1}^{w} \Pr(\bar{b}_i|\Phi)) \qquad (8.15)$$

In terms of specific implementation of Φ, we adopt the Variable order Markov Model (VMM) [2], which is a category of algorithms for prediction of discrete sequences. It can capture both large and small order Markov dependencies extracted from $\overline{D}_A$. Therefore, it can encode richer and more flexible Sequential Cue. This can be done efficiently by constructing a probability suffix tree (PST) [22, 30], a fast implementation algorithm of VMM. Algorithm 2 shows the detail of this process.

4.3 Final Early Classification Function

Given an ongoing m.t.s. Y', we can now construct our final early classification function $\mathcal{F}(Y')$ by using the knowledge learned from Sects. 4.1 and 4.2, namely *time dynamics* model $\Pr(Y'|\Lambda)$ and *sequential cue* model $\Pr(Y'|\Phi)$. We use MD-MPP to denote the first model, and MD-MPP+TD to denote the second model. The early classification performances are evaluated on both of the two models. For a domain application with $|C|$ classes, our final prediction functions of two models can be written as:

$$\text{MD-MPP:} \qquad \mathcal{F}(Y') = \arg\max_{c\in C}\{\Pr^c(Y'|\Lambda)\};$$

$$\text{MD-MPP+TD:} \qquad \mathcal{F}(Y') = \arg\max_{c\in C}\{\Pr^c(Y'|\Phi)\}.$$

5 Experimental Studies

In this section, we present a comprehensive evaluation of our methods (MD-MPP and MD-MPP+TD) on modeling accuracy and time efficiency using two real-world human activity data sets. We have compared with the state-of-the-art methods including 1NN+DTW [16], ECTS [37], MSD [9], and HMM.[4] Table 8.2 summarizes the baselines.

---

[4]We used the following public available toolbox as our HMM implementation: http://www.cs.ubc.ca/~murphyk/Software/HMM/hmm.html.

Table 8.2 Summary of the four baselines used for quantitative comparison with our algorithm

Methods	Rationale	Description
1NN+DTW [16]	State-of-the-art	The dynamic time warping (DTW) based distance measurements between test and training time series are computed for use in 1NN classifier. For m.t.s., the distance is measured as the average of component u.t.s. distances
ECTS [37]	Extension of 1NN	The MPL (Minimum Prediction Length) for a cluster of similar time series are computed first. At the testing phase, the learned MPLs are used to select the nearest neighbor from only "qualified" candidates in terms of MPL. For m.t.s., the distance is measured as the average of component u.t.s. distances
MSD [9]	Extension of shapelets	Multivariate shapelets are extracted using a sliding-window based strategy. These shapelets are then pruned according to the weighted information gain
HMM	Common statistical model	The HMM is selected as a representative of generative model based methods. A model is trained for each class. Decisions are based on likelihoods ranking

5.1 Datasets

We utilized two real-world datasets: CMU human motion capture dataset [5], UCI Australian Sign Language (Auslan) dataset [25]. The following details the collection and preprocessing of the two datasets.

Human Action Data The dataset was composed of dozens of synchronized motion capture actions performed by more than one hundred subjects. In our experiment, we select the MoCap data of nine common action classes performed by different subjects, which consists of 10 samples per class on average (total 91 samples) with average duration of 839 frames. The nine action classes include *run*, *pick-up*, *walk*, *jump*, *sitting on a motorcycle*, *boxing*, *cartwheel*, *chicken dance*, and *golf swing*. The human body model consists of 34 bones with hierarchical structures. Each action is specified by m.t.s. observations on motion angles of body bones, which describe both moving direction and magnitude of joints, as well as the dynamic relationships between bones. Figure 8.3 shows the human body model with the above-mentioned hierarchical structure. The original full body Degree of Freedoms (DOFs) are 62. We discard some unimportant joint angles, such as fingers, toes, thumb in the experiments. Finally, we select 19 body joints which cover the DoFs of radius, humerus, tibia, femur, and the upper back.

Sign Language Data This dataset was composed of sample of Auslan (Australian Sign Language) signs [13, 24], in which 27 samples of each of 95 Auslan signs (in total 2565 signs) were captured from a native signer using high-quality position trackers (two Fifth Dimension Technologies gloves). Each hand has 11 degrees of freedom (i.e., roll, pitch and yaw as well as x, y and z), which results in a total of 22 dimensional m.t.s. observations of signs. The average length of each sign is approximately 57 frames, where the refresh rate is close to 100 frames per second.

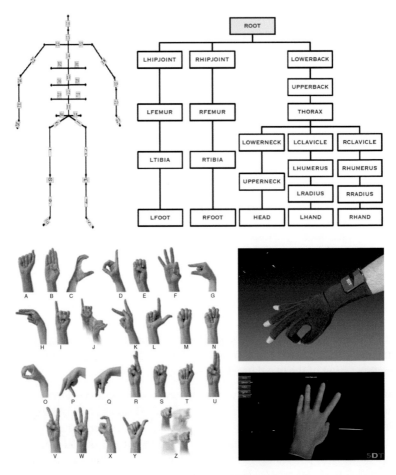

Fig. 8.3 Evaluation datasets. *Top*: CMU pose. *Bottom*: UCI sign language

5.2   Performance Comparison

We compare our algorithms of m.t.s. early classification in Sect. 4 (MD-MPP and MD-MPP+TD) with existing alternatives that we discussed in Sect. 2 and summarized in Table 8.2. We evaluate the classification accuracy by using the standard "leave-one-out" method in both two datasets. Different from traditional classification task, for early classification, we focus on the predictive power of each method. An early classifier should use an observation ratio as small as possible to make an accurate prediction. To evaluate this, we do classification along the progress of time series, and predict the class label at different progress levels (observation ratio) of time series. Specifically, we use 5 % of full m.t.s. duration as an interval of generating a new prediction result.

Fig. 8.4 Performance comparisons on two datasets (see text for detailed discussions). In each figure, the vertical axis is the classification accuracy averaged over all classes, and the horizontal axis is the observation ratio, which can be viewed as the normalized time line $((0, T] \rightarrow (0, 1])$. (**a**) CMU human action dataset (**b** UCI sign language dataset)

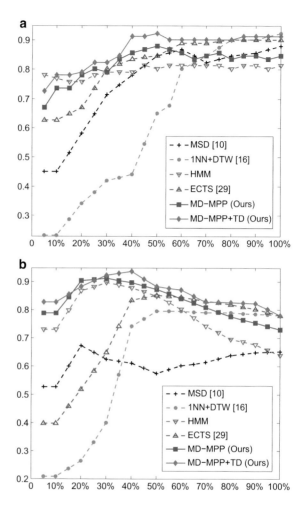

Model Construction For the Human Action Data, we construct an MD-MPP model by splitting mark space into 10 levels ($L = 10$) and dividing the time line into 20 pieces of equal length segments ($S = 20$). To construct an MD-MPP+TD model, we train an order 3-bounded PST ($O = 3$) first, then do 100 times sampling ($w = 100$) of event streams for each m.t.s. at testing phase. For the Sign Language Data, we set $L = 20$, $S = 10$, $O = 3$, and $w = 100$.

Results Figure 8.4 summarizes the quantitative comparison between our methods and four baselines. These graphs help us make the following observations:

(1) Our algorithms significantly outperform all the compared methods in most cases, and achieve high prediction accuracy over different levels of observation ratio. In terms of full-length classification (at observation ratio 100 %), 1NN-DTW is the most comparable one to ours, which demonstrates its robustness

Table 8.3 Performance comparisons on two datasets (percentage as observation ratios)

Methods	CMU pose dataset					UCI signLanguage dataset				
	20,%	40 %	60 %	80 %	100 %	20 %	40 %	60 %	80 %	100 %
1NN+DTW [16]	0.34	0.44	0.80	0.90	**0.92**	0.26	0.74	0.80	0.79	0.78
ECTS [37]	0.67	0.84	0.89	0.90	0.90	0.52	0.84	0.85	0.82	0.78
MSD [9]	0.58	0.78	0.87	0.85	0.88	0.67	0.61	0.60	0.64	0.65
HMM	0.76	0.79	0.81	0.80	0.81	0.87	0.88	0.80	0.70	0.64
MD-MPP (Ours)	0.78	0.86	0.86	0.84	0.85	**0.90**	0.90	0.84	0.78	0.73
MD-MPP+TD (Ours)	**0.79**	**0.91**	**0.90**	**0.90**	0.91	0.88	**0.94**	**0.87**	**0.83**	**0.78**

The bold values indicate the best result (i.e., highest accuracy) in each setting.

as the state-of-the-art method for time series classification. At early stages
of observation ($< 30\,\%$), MSD and ECTS can outperform 1NN-DTW to
accomplish better early classification due to their designs on utilizing early
cues. As a latent state model, HMM is relatively less dependent on full length
observation. Table 8.3 shows detailed comparisons of six methods on two
datasets.

(2) Each dataset has different predictability, which means the discriminative segments of m.t.s. may appear at different stages of time series. As illustrated
in Fig. 8.4, we achieved near-optimal classification accuracy at the observation
ratio of 40 % in the Human Action Data, and 20 % in the Sign Language Data,
respectively. Figure 8.5 shows the corresponding detailed results in confusion
matrices, respectively. Interestingly, Fig. 8.4b shows that the prediction accuracy does not necessarily go up with the increasing of information observed,
which means more noise is introduced at the late stages of m.t.s.. It is probably
because that different signs end in the same way, such as open palms or fists.

(3) Sequential cue patterns are contributing for better prediction, as shown
in Action and Sign Language datasets (Fig. 8.4a,b). This is because
variables/events have strong correlations (for example, bones connected to
each other) in Action and Sign Language datasets.

(4) In Fig. 8.6, we present detailed performance of our approach over nine different
action classes in the Human Action dataset. The action "pick-up" is difficult
to be recognized at early stages, because it is executed by first walking to the
object, then picking up it. The component sub-action "walking to object" makes
it confusing with the class "walk." Another component sub-action "crouching
to pick up object" makes it confusing with the class "jump."

6 Summary

Action recognition is an important research study of Human Motion Analysis
(HMA). This work takes one step further, focusing on early recognition of ongoing

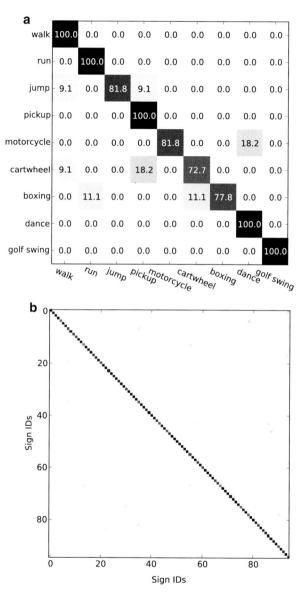

Fig. 8.5 Confusion matrices on two datasets. (**a**) CMU dataset (at **40** % observation ratio) (**b**) UCI dataset (at **20** % observation ratio)

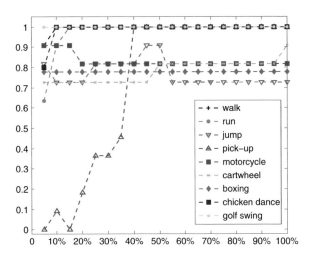

Fig. 8.6 Detailed results on Human Action dataset

human actions, which is beneficial for a large variety of time-critical applications, e.g. gesture based human machine interaction and somatosensory game, etc. Our goal is to identify class label of human activities with partial observation of temporally incomplete action executions. By considering action data as multivariate time series (m.t.s.) synchronized to a shared common clock (frames), we introduced a novel approach to early classify human activities (m.t.s. observation) by modeling two types of time pattern: *temporal dynamics* and *Sequential Cue*. The major contributions include a Multilevel-Discretized Marked Point Process (MD-MPP) model for representing m.t.s.; and a sequential cue model (MD-MPP+TD) to characterize the sequential patterns among multiple time series variables. We have empirically shown that our approach is superior in the activity prediction task.

References

1. Aggarwal, J.K., Ryoo, M.S.: Human activity analysis: A review. ACM Comput. Surv. (CSUR) **43**(3), 16 (2011)
2. Begleiter, R., El-Yaniv, R., Yona, G.: On prediction using variable order markov models. J. Artif. Intell. Res. **22**, 385–421 (2004)
3. Bregón, A., Aránzazu Simón, M., José Rodríguez, J., et al.: Early fault classification in dynamic systems using case-based reasoning. In: Current Topics in Artificial Intelligence, pp. 211–220. Springer, New York (2006)
4. Cao, Y., Barrett, D., et al.: Recognizing human activities from partially observed videos. In: Proceeding IEEE Conference Computer Vision and Pattern Recognition, IEEE, New York (2013)
5. CMU: http://mocap.cs.cmu.edu
6. Daley, D.J., Vere-Jones, D.: An Introduction to the Theory of Point Processes, Vol. I, vol. 1. Springer, New York (2003)

7. Eruhimov, V., Martyanov, V., Tuv, E.: Constructing high dimensional feature space for time series classification. In: Knowledge Discovery in Databases: PKDD, pp. 414–421. Springer, New York (2007)

8. Fu, T.C.: A review on time series data mining. Eng. Appl. Artif. Intell. **24**(1), 164–181 (2011)

9. Ghalwash, M., Obradovic, Z.: Early classification of multivariate temporal observations by extraction of interpretable shapelets. BMC Bioinf. **13**(1), 195 (2012)

10. Gunawardana, A., Meek, C., Xu, P.: A model for temporal dependencies in event streams. In: The Annual Conference on Neural Information Processing Systems, pp. 1962–1970 (2011)

11. Hadfield, S., Bowden, R.: Hollywood 3d: Recognizing actions in 3d natural scenes. In: IEEE Conference on Computer Vision and Pattern Recognition (CVPR), pp. 3398–3405 (2013)

12. Hoai, M., De la Torre, F.: Max-margin early event detectors. In: Proceeding of IEEE Conference Computer Vision and Pattern Recognition, pp. 2863–2870 (2012)

13. Ibarguren, A., Maurtua, I., Sierra, B.: Layered architecture for real-time sign recognition. Comput. J. **53**(8), 1169–1183 (2010)

14. Jansen, A., Niyogi, P.: Point process models for event-based speech recognition. Speech Comm. **51**(12), 1155–1168 (2009)

15. Katagiri, H., Nishizaki, I., Hayashida, T., et al.: Multiobjective evolutionary optimization of training and topology of recurrent neural networks for time-series prediction. Comput. J. **55**(3), 325–336 (2012)

16. Keogh, E.: Exact indexing of dynamic time warping. In: Proceedings of the 28th International Conference on Very Large Data Bases, pp. 406–417. VLDB Endowment (2002)

17. Keogh, E., Kasetty, S.: On the need for time series data mining benchmarks: a survey and empirical demonstration. In: ACM SIGKDD Conference on Knowledge Discovery and Data Mining, pp. 102–111. ACM, New York (2002)

18. Kim, G., Fei-Fei, L., Xing, E.P.: Web image prediction using multivariate point processes. In: ACM SIGKDD Conference on Knowledge Discovery and Data Mining, pp. 1068–1076. ACM, New York (2012)

19. Kong, Y., Kit, D., Fu, Y.: A discriminative model with multiple temporal scales for action prediction. In: European Conference on Computer Vision, pp. 596–611. Springer, New York (2014)

20. Koppula, H., Saxena, A.: Learning spatio-temporal structure from rgb-d videos for human activity detection and anticipation. In: Proceedings of the 30th International Conference on Machine Learning, pp. 792–800 (2013)

21. Li, K., Fu, Y.: Prediction of human activity by discovering temporal sequence patterns. IEEE Trans. Pattern Anal. Mach. Intell. **36**(8), 1644–1657 (2014)

22. Li, K., Hu, J., Fu, Y.: Modeling complex temporal composition of actionlets for activity prediction. In: European Conference on Computer Vision, pp. 286–299. Springer, New York (2012)

23. Li, K., Li, S., Fu, Y.: Early classification of ongoing observation. In: 2014 IEEE International Conference on Data Mining (ICDM), pp. 310–319 (2014)

24. Lichtenauer, J.F., Hendriks, E.A., Reinders, M.J.: Sign language recognition by combining statistical dtw and independent classification. IEEE Trans. Pattern Anal. Mach. Intell. **30**(11), 2040–2046 (2008)

25. Lichman, M.: UCI Machine Learning Repository. School of Information and Computer Science, University of California, Irvine. http://archive.ics.uci.edu/ml (2013)

26. Liu, L., Shao, L.: Learning discriminative representations from rgb-d video data. In: Proceedings of the Twenty-Third International Joint Conference on Artificial Intelligence, pp. 1493–1500. AAAI Press (2013)

27. Luo, J., Wang, W., Qi, H.: Group sparsity and geometry constrained dictionary learning for action recognition from depth maps. In: IEEE International Conference on Computer Vision, pp. 1809–1816 (2013)

28. Prabhakar, K., Oh, S., Wang, P., et al.: Temporal causality for the analysis of visual events. In: CVPR, pp. 1967–1974. IEEE, New York (2010)

29. Rodríguez, J.J., Alonso, C.J., Boström, H.: Boosting interval based literals. Intell. Data Anal. 5(3), 245–262 (2001)
30. Ron, D., Singer, Y., Tishby, N.: The power of amnesia: Learning probabilistic automata with variable memory length. Mach. Learn. 25(2), 117–149 (1996)
31. Ryoo, M.S.: Human activity prediction: Early recognition of ongoing activities from streaming videos. In: Proceedings of IEEE Int'l Conference Computer Vision, pp. 1036–1043 (2011)
32. Wang, J., Liu, Z., Wu, Y., Yuan, J.: Mining actionlet ensemble for action recognition with depth cameras. In: 2012 IEEE Conference on Computer Vision and Pattern Recognition (CVPR), pp. 1290–1297. IEEE, New York (2012)
33. Wei, L., Keogh, E.: Semi-supervised time series classification. In: ACM SIGKDD Conference on Knowledge Discovery and Data Mining, pp. 748–753. ACM, New York (2006)
34. Xi, X., Keogh, E., Shelton, C., et al.: Fast time series classification using numerosity reduction. In: International Conference on Machine Learning, pp. 1033–1040. ACM, New York (2006)
35. Xia, L., Aggarwal, J.K.: Spatio-temporal depth cuboid similarity feature for activity recognition using depth camera. In: IEEE Conference on Computer Vision and Pattern Recognition (CVPR), pp. 2834–2841 (2013)
36. Xing, Z., Pei, J., Yu, P., Wang, K.E.: Extracting interpretable features for early classification on time series. In: SIAM International Conference on Data Mining (2011)
37. Xing, Z., Pei, J., Yu, P.S.: Early prediction on time series: a nearest neighbor approach. In: International Joint Conference on Artificial Intelligence, pp. 1297–1302 (2009)
38. Ye, L., Keogh, E.: Time series shapelets: a new primitive for data mining. In: ACM SIGKDD Conference on Knowledge Discovery and Data Mining, pp. 947–956. ACM, New York (2009)
39. Zhang, Z., Cheng, J., Li, J., et al.: Segment-based features for time series classification. Comput. J. 55(9), 1088–1102 (2012)

Printed in the United States
By Bookmasters